LASERS AND ELECTRO-OPTICS RESEARCH AND TECHNOLOGY

PHASE-MATCHED HIGH ORDER HARMONIC GENERATION AND APPLICATION

DINH BA KHUONG
AND
DAO VAN LAP

nova publishers
New York

Library of Congress Cataloging-in-Publication Data

ISBN: 978-1-62618-128-1

Published by Nova Science Publishers, Inc. † New York

CONTENTS

SUMMARY

In May 1960 Theodore Maiman demonstrated the first laser and this opened up the remarkable field of nonlinear optics which has been developing rapidly. An intense laser field allows electrons to be strongly perturbed and thus nonlinear phenomena can occur. For example, the second optical harmonic in a crystal was first observed in 1961 [1] and the third harmonic in a rare gas was first reported in 1967 [2].

High Harmonic Generation (HHG) is a highly nonlinear process in which harmonics are produced by an intense laser field when it interacts with an atomic or molecular gas [3, 4]. This phenomenon was first observed in 1987 by McPherson et al. who successfully generated harmonic emission up to the 17th order at 248 nm in neon gas [5]. The typical intensity required for generation of high order harmonics in the noble gases is of the order of 10^{13}-10^{14} W/cm^2 [6, 7] and can be achieved by focusing an amplified femtosecond laser beam. In 1992, Krause et al. [3] and in 1993 Corkum et al. [8] proposed a theoretical explanation for generation of high order harmonics in terms of a three step model. In this model, free electrons produced by ionization by the laser field and then accelerated by the laser field recombine with their parent ions, releasing energy as single high energy photons. The output HHG flux is a combination of the single atom response and the macroscopic propagation through the medium [6, 7]. HHG produces a coherent radiation source in the extreme ultraviolet and soft x-ray regions of the spectrum [9, 10]. In addition, with the possibility of generating very short pulses of duration of the order of tens of attoseconds, i.e., the natural time scale of atomic physics, HHG sources are very useful in a wide range of practical applications such as time-resolved studies of ultrafast dynamics in atoms [11], molecular dynamics [12], solid state science [13], and plasma diagnostics [14]. Moreover, short wavelength light in the extreme ultraviolet and soft x-ray region with high coherence and high flux produced by HHG can be employed in a lensless imaging technique known as coherent diffractive imaging (CDI) [15]. By recording a far-field coherent diffraction pattern and using a phase retrieval algorithm, this technique eliminates all optical elements such as lenses and is therefore free from aberration and provides a very large depth of focus to meet demands for nondestructive imaging applications requiring high resolution in thick samples [15]. Numerous core-shell absorption edges and widely varying elemental absorption cross sections provide excellent image contrast, in particular for imaging biological molecules such as membrane proteins in the water window region of the spectrum (~ 2.3 nm to 4.4 nm) [10, 16].

In this book, we consider experimental work on the generation of high order harmonics and use it as a source for some applications. The structure of the book is as follows. Chapter 1 is a literature review on the HHG technique. This chapter begins by summarizing the history of HHG. The principles of the HHG emission are then discussed. This is followed by a discussion of the macroscopic phase matching effects which influence the coherent construction of the harmonic emission. The enhancement of phase matching and the quasi phase matching in HHG are also briefly described in the next section. The chapter finishes with a description of the high power femtosecond laser system and the fundamental experimental arrangements for generation and detection of the harmonic radiation in a semi-infinite gas cell which is an important configuration of our experimental setup.

Chapter 2 presents two new schemes for enhancement of the harmonic generation. In the first part of the chapter, by using a combination of axicon and lens to influence the geometrical phase mismatch leading to an improvement of the phase matching condition, it is shown that for the harmonic emission in the cut-off region the photon flux and the spatial beam profile are strongly enhanced. In the second part, an off-axis laser beam is used to control the phase matching condition in the HHG process and thus the harmonic signal can be suppressed or enhanced. High photon energy up to 220 eV (~ 5.5 nm in wavelength) from helium gas is achieved in our experiments in the presence of the off-axis beam. The important role of the harmonic dipole phase on the phase-matched HHG process, which has not been considered in most previous studies, is also considered in this section.

Chapter 3 investigates the influence of the driving laser on the spectral features of the harmonic generation. The change of the harmonic spectrum when varying the phase of the driving field by controlling the frequency chirp of the laser is analyzed and this allows us to properly understand the role of the phase of the fundamental field on the HHG process. By using a combination of a half-wave plate and a polarizing beam splitter to control the laser pulse energy precisely and continuously without any change of the focusing geometry, the laser intensity dependence of the harmonic generation is discussed. Based on this investigation, the effects of the harmonic dipole phase and the dispersion phase mismatches induced by the ionized medium during the harmonic generation process are clearly revealed and the interplay between the macroscopic response and the single-atom response is also shown. Moreover, we discuss that the use of an aperture to limit the ionization to less than a critical level and to control the wave-front of the fundamental field are required for sharp and strong harmonic generation.

The next two chapters consider applications of the high harmonic generation. Chapter 4 proposes a new method to obtain information about molecular structure dynamics based on phase matched optimized harmonic generation in a semi-infinitive gas cell. The chapter begins with an overview of HHG from diatomic molecules. The influence of field-free alignment of the molecules on the phase matching conditions is then studied. The experimental setup and discussion of results relating to periodic modulations in the harmonic intensity are then presented.

Chapter 5 is devoted to the topic of coherent diffractive imaging (CDI) in the extreme ultraviolet region. We briefly review the theoretical background of this imaging technique including an introduction to CDI, phase retrieval and the experimental requirements. In the following section we show experimental results on CDI using a high order harmonic source around 13.5 nm based on a helium semi-infinite gas cell for two samples: a 2D regular array of pinholes and a conjugated polymer film. The generation of the illumination source and the spatial coherence studies based on a Young's double slit experiment are presented. Imaging results are shown in the following section of this chapter.

Chapter 1

HIGH HARMONIC GENERATION

1.1. OVERVIEW OF HIGH HARMONIC GENERATION

1.1.1. History of High Harmonic Generation

High Harmonic Generation (HHG) occurs when an intense laser pulse is focused into a noble gas [3, 4]. HHG provides an attractive source of coherent radiation in the extreme ultraviolet (XUV) radiation and soft X-ray region and is currently finding wide application in physics, chemistry, biochemistry and biology [17].

Harmonic Generation was first observed in 1961 by P.A. Franken et al. at the University of Michigan [1] soon after Theodore Maiman demonstrated the first laser in May 1960. In the experiment, Franken and his colleagues focused light from a pulsed ruby laser (at 694.3 nm wavelength, ~ 3 J pulse energy and 1 ms pulse duration) into a non isotropic crystalline quartz sample. As a consequence, the second harmonic of the fundamental beam at 347.2 nm was generated with a very low efficiency conversion. This is known as the first experiment and breakthrough in the field of non-linear optics.

In 1967, New et al. demonstrated the first third-harmonic generation in a gas [2]. However, for many years, only low harmonics (second up to fifth order) were studied, until in the late 1980s. In 1987, by interacting krypton fluoride excimer laser pulses (at 248 nm, ~20 mJ pulse energy and 350 fs pulse duration) with neon gas, McPherson et al. at the University of Illinois in Chicago successfully generated harmonic emission up to the 17th order [5]. From then onwards, many other researchers have studied harmonic generation to obtain the highest possible order and conversion efficiency. For example,

L'Huillier et al. [18] observed the 29th harmonic in Xe, the 57th harmonic in Ar, and the 135th harmonic in Ne by using a Nd:glass laser with a wavelength of 1053 nm and pulse duration of 1 ps.

1.1.2. Principles of High Harmonic Generation

1.1.2.1. Extreme Nonlinear Optics

Following the invention of the laser in 1960, the science of nonlinear optics rapidly developed and has become one of the most significant technologies of the 20th century. This is opposite to traditional optics theory in which a linear relationship between the low applied electric field E of light and the dielectric polarization P of a material is given by

$$P = \varepsilon_0 \chi E \tag{1.1}$$

where χ is the linear susceptibility and ε_0 is the permittivity.

However, at strong electric fields (up to 10^8 V/m), i.e., when the laser field becomes comparable to the internal atomic Coulomb field, the nonlinear terms in the induced polarization become important. The polarization can be expressed by the Taylor series

$$P = \varepsilon_0 \chi^{(1)} E + \varepsilon_0 \chi^{(2)} E^2 + \varepsilon_0 \chi^{(3)} E^3 + \dots \tag{1.2}$$

The first term defines the linear susceptibility, the second term defines the lowest order nonlinear susceptibility, and so on. The optical nonlinearities are small compared with the first term. When a very strong laser field is applied, a very high order of the polarization is induced. Under certain conditions, the nonlinear polarization can act as a source for the generation of frequency up-conversion of the fundamental field.

High Harmonic Generation is an extreme nonlinear optical process which occurs in the strong field regime. The required focused intensity of the fundamental beam for this process is at least 10^{13} W/cm^2 and can be obtained by focusing a high power femtosecond laser beam. At this laser intensity, the contribution of very high order nonlinearities becomes significant. However, it is worth noting that there is an upper limit of laser intensity for HHG. This is related to optical breakdown when the plasma electron density reaches a critical density and becomes opaque to the pumping laser radiation. In addition, at an intensity ~ 10^{16} W/cm^2 the increased magnetic field prevents recombination with the parent atom, which suppresses the HHG process. Self-focusing of the laser beam and the creation of plasma are also limiting factors at high intensity.

Depending on the medium, such as a gas of atoms, molecules, or a solid material, the necessary focused intensities are different and the response of the medium to the laser field is also different.

1.1.2.2. Three Step Model

Since the early discovery of High Harmonic Generation, there have been many theoretical efforts to explain the mechanism of the HHG process. To describe accurately the HHG process, the fully numerical solution of the time-dependent Schrödinger equation (TDSE) needs to be calculated. However, a quasi classical three step model can clearly provide a physical picture of the interaction between atoms and strong laser fields [3, 8] and can be used to predict the features of HHG. In this model, under the interaction of a strong laser field, the active electrons first tunnel through the potential barrier, are then accelerated, and when the laser field changes sign the accelerated electrons are pulled back and finally recombine with parent ions to emit high-energy photons.

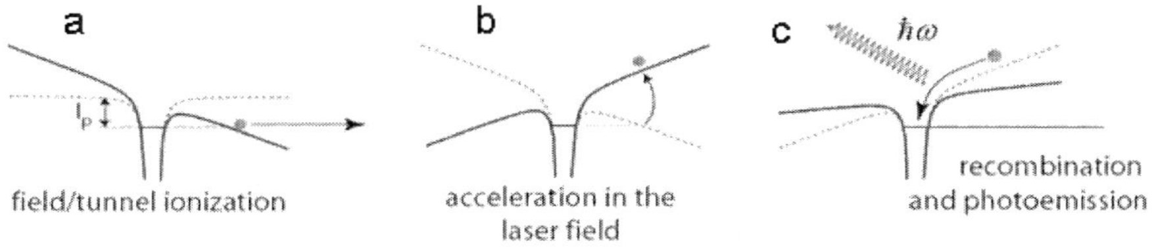

Figure 1.1. Three step model of High Harmonic Generation.

In the first step, the intense laser field modifies the Coulomb potential of the atom, allowing the electron to tunnel through the barrier and ionize (figure 1.1a). After the tunnelling process, the electron is considered as a "free particle" in the electric field. The free electron accelerates in the laser field and gains momentum (figure 1.1b). When the field reverses, the electron is accelerated back toward the ionic parent. If recombination occurs, the electron releases a photon with very high energy (figure 1.1c).

The fraction of ionization in the tunnelling regime was first studied by Ammosov, Delone and Krainov (ADK model) [20]. The rate of ionization from the ground state can be expressed as

$$\omega(t) = \omega_P \left|C_{n^*}\right|^2 \left(\frac{4\omega_P}{\omega_t}\right)^{2n^*-1} \exp\left(-\frac{4\omega_P}{3\omega_t}\right) \tag{1.3}$$

where

$$\omega_P = \frac{I_P}{\hbar}$$

$$\omega_t = \left(\frac{eE(t)}{2m_e I_P}\right)^{1/2}$$

$$n^* = Z\left(\frac{I_H}{I_P}\right)^{1/2}$$

$$\left|C_{n^*}\right|^2 = 2^{2n^*}\left[n^*\Gamma(n^*+1)\Gamma(n^*)\right]^{-1}$$

in which I_P is the ionization potential of the atom, I_H is the ionization potential of atomic hydrogen, E(t) is the electric field of the laser, m_e is the electron mass, Z is the ion charge after ionization, and Γ is the Gamma function.

The fraction of ionization is given by

$$\eta(t) = \exp\left[-\int_{-\infty}^{t}\omega(t')dt'\right] \tag{1.4}$$

During the second step, the electron gains mean kinetic energy known as the ponderomotive energy U_P. This energy is directly proportional to the intensity I of the incident laser and the square of the fundamental wavelength λ. It is given by

$$U_P = \frac{e^2E^2}{4m_e\omega_0{}^2} = \frac{e^2I}{2m_e\omega_0^2\varepsilon_0 c} \propto I\lambda^2 \tag{1.5}$$

where e is the electron charge, E is the electric field and ω_0 is the angular frequency of the driving laser, respectively.

When the electron returns to its parent ion, the recollision energy is not only the kinetic energy (U_P) but also the ionization potential of the atom (I_P). Thus, according to this model, the maximum energy (cutoff energy) is given by [3, 8]

$$E_{cutoff} = \hbar\omega_{max} = I_P + 3.17U_P \tag{1.6}$$

where ω_{max} is the maximum angular frequency and $\hbar$ is the reduced Planck constant.

We know that $\omega_{max} = q_{max}\omega_0$, where q_{max} is the cutoff harmonic order.

Hence, the cutoff harmonic order is

$$q_{max} = \frac{I_P + 3.17U_P}{\hbar\omega_0} \tag{1.7}$$

The term $3.17U_P$ is the maximum kinetic energy upon the electron returning to the nucleus and this happens for the phase of the electric field at the instant of ionization, $\varphi \approx 17^0$ [19], which is close to the maximum of the electric field.

According to equation (1.6), the cutoff can be extended by increasing the ponderomative energy, U_p. This can be achieved by using a longer wavelength of the fundamental laser or a higher laser intensity. In addition, based on equation (1.6), the extension of the position of the cutoff is also possible by using atoms with high ionization potentials, I_p, like the noble gases. For example, if a fundamental beam with wavelength of 800 nm and duration of 26 fs is applied, the maximum harmonic orders that can be generated in the noble gases are H27, H41, H61, H155, H221 for Xe, Kr, Ar, Ne, and He, respectively.

1.1.2.3. Typical HHG Spectrum

As mentioned in the three step model, at one half-cycle after ionization, the electron turns back towards the parent ion to recombine when the laser electric field changes sign. Hence, the harmonics are produced twice each cycle and each half cycle of the driving laser produces a short (sub-femtosecond) burst of XUV radiation. Thus, the harmonics are characterized by a series of bursts in the time domain separated by half the laser period.

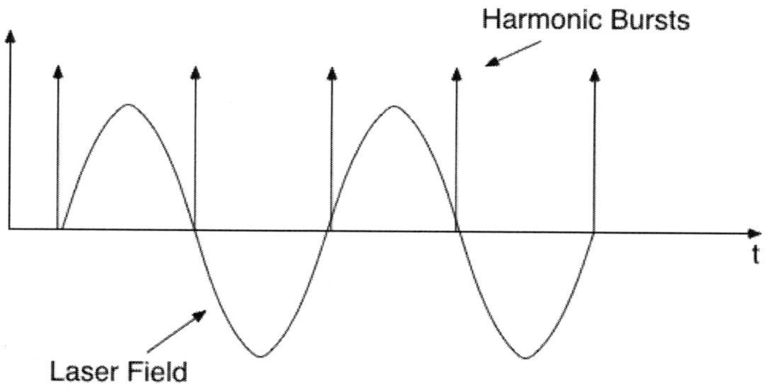

Figure 1.2. Harmonic bursts in time domain.

By taking a Fourier transform, this separation of half the laser period in the time domain results in peaks separated by $2\omega_0$ in the frequency domain. In addition, since the consecutive bursts are a consequence of collisions from opposite directions, the corresponding spectral components have the same amplitude but opposite in phase. This leads to destructive spectral interference for even order harmonics ($\omega = 2m\omega_0$), and constructive interference for odd harmonics ($\omega = (2m+1)\omega_0$). This is why, normally, only odd harmonics are observed in the HHG spectrum.

A typical HHG spectrum can be divided into three parts: the perturbative region at low orders, the plateau region at intermediate orders and the cutoff region at highest orders. Perturbation theory can be used to describe the appearance of lower order harmonics ($q < 9$) which are produced at low intensities during the leading edge or trailing edge of the laser pulse. Basically, the harmonic yields in this region decrease as a power law.

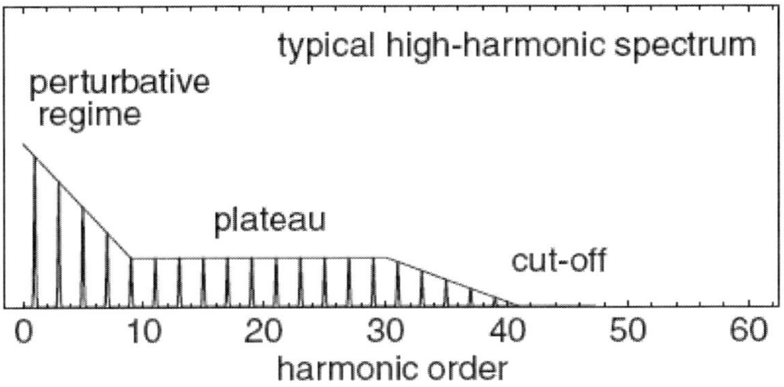

Figure 1.3.Typical HHG spectrum.

The higher order harmonics are generated at higher laser intensities. When the electric field strength of the laser is on the same scale as the inner atomic electric fields, it cannot be considered as a small perturbation anymore, i.e., perturbation theory breaks down. The plateau region contains the harmonics which are produced when applying a high strength driving electric field. In this region, the harmonic intensities are relatively unchanged with increasing order. Another interesting point is that, in the plateau region, two electron trajectories contribute to the harmonic spectrum. Electrons in these two trajectories are ionized at different phases of the laser field. However, they can return to the core with the same velocity, i.e., the same final kinetic energy at recombination, leading to the emission of radiation at the same frequency. An electron on the long trajectory which is ionized at a phase of 3^0 returns to the parent ion close to one period of an optical cycle while an electron on the short trajectory which is produced at a phase of 45^0 has a return time less than half an optical cycle [19].

The plateau region is followed by the cutoff region. The cutoff photon energy is determined by the maximum energy the recombining electron can obtain in the electric field given by equation (1.6). This occurs for an electron which is ionized at a phase of the laser field of 17^0 and therefore the maximum kinetic energy of 3.17Up is achieved.

1.1.2.4. Strong Field Approximation (SFA) Model (Quantum Model)

The three step model provides the basic picture of the HHG process and has been successful in explaining the experimental observations, like the prediction of the cutoff region. However, it fails to explain the spectral characteristics of the plateau region.

To deal with this problem, and to understand the HHG process more precisely, a fully quantum mechanical theory has been developed within the strong field approximation by Lewenstein et al. [21]. In this model, Lewenstein et al. solved numerically the time dependent Schrödinger equation in a strong field with the assumptions:

(a) The contribution to the wave packet of all bound states except the ground state can be neglected
(b) There is no depletion of the ground state
(c) In the continuum states, the electron can be treated as a free particle moving in the electric field with no effect of the atomic potential

Lewenstein et al. claimed that an electron is promoted into a continuum state by the laser field, then propagates in the laser field, and finally recombines with the parent ion by a dipole transition. In addition, in the SFA approach, calculation of the harmonic dipole moment is performed as the coherent sum of all the different quantum paths which contribute to the harmonic emission. These quantum paths are a generalization of the classical electron trajectories as defined in section 1.1.2.3. The phase associated with each quantum path is given by the classical action accumulated by the electron along the corresponding trajectory and can be approximated by the product of the ponderomative energy and the electron excursion time.

Based on the SFA model, it is shown that the phase of the dipole moment depends on the laser intensity. This leads to an intensity dependent harmonic phase which helps us to explain successfully different aspects of the generation process. Also, the Lewenstein model predicts a cutoff energy, $E_{cutoff} = \hbar\omega_{max} = FI_P + 3.17U_P$, slightly higher than that in the three step model. The factor $F(I_P / U_P)$ is equal to 1.3 for $I_P << U_P$ and approaches unity as I_P grows.

1.1.3. Phase Matching in High Harmonic Generation

1.1.3.1. Definition of Phase Matching

Studying the HHG process in a single atom response basically provides the predicted shape of the harmonic spectrum. However, it cannot account for the observed harmonic yield. To obtain strong harmonic emission in a medium, propagation effects need to be considered. So far, there has been much work in which these problems have been investigated [6, 7, 17, 22-25]. For coherent construction of the harmonic emission, the wave fronts of the fundamental laser and the generated harmonics must be in phase. Atoms along the path of the laser produce harmonics propagating with the laser field. When they are in phase with the laser wave front, there is constructive interference and these harmonics combine to build up the harmonic emission.

The distance over which the fundamental wave front and the harmonic wave front become out of phase is called the coherence length and is given by

$$L_{coh} = \frac{\pi}{\Delta k} \tag{1.8}$$

where Δk is the phase mismatch.

For phase matching in the HHG process, one needs $\Delta k_q = qk_0 - k_q = 0$, where q is the harmonic order, k_q is the wavevector of the qth harmonic and k_0 is the wave vector of the fundamental beam.

In general, the wavevector k of light with wavelength λ propagating in a gas medium is given by [19]

$$k = \frac{2\pi}{\lambda} + \frac{2\pi N_a n(\lambda)}{\lambda} - N_e r_e \lambda \tag{1.9}$$

where N_a is the atom density, N_e is the free electron density in the medium, n(λ) is the refractive index per unit neutral atom density at wavelength λ, and r_e is the classical electron radius.

In equation (1.9), on the right hand side, the first term is due to the dispersion of the vacuum, the second results from the dispersion of the gas and the third comes from the dispersion of the induced plasma.

The intensity I_q of the qth harmonic at the end of the nonlinear medium without absorption is given in [26]

$$I_q \propto L^2 \sin^2(\Delta k_q L/2)/(\Delta k_q L/2)^2 \tag{1.10}$$

where L is the length of the medium.

From equation (1.10), when the phase mismatch is zero, the harmonic intensity increases quadratically with the propagation distance. Otherwise, the harmonic intensity will oscillate sinusoidally with propagation distance and the periodicity of this oscillation is twice the coherence length, L_{coh}.

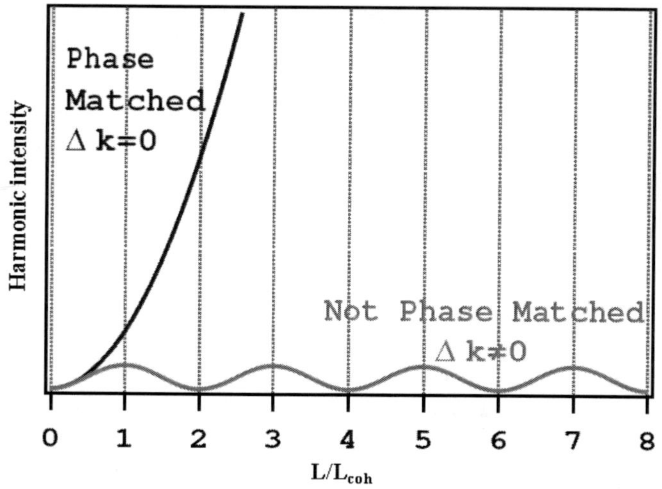

Figure 1.4.Harmonic intensity as a function of propagation distance L.

1.1.3.2. Phase Mismatch

The harmonic yield is severely limited by phase mismatch phenomena, such as the difference between the diffraction rates for the fundamental beam and the individual harmonic radiation due to geometrical propagation effects, the wavelength dependent index of refraction of the neutral atomic and ionized medium, and the dependence of the intrinsic phase of the harmonics on the laser intensity in both the longitudinal and radial directions. In general, the total phase mismatch is the sum of four terms and can be expressed as [27, 28]

$$\Delta k_q = qk_0 - k_q = \frac{2\pi q}{\lambda} p\delta n(1-\eta) - \eta p N_{atm} r_e \lambda (q - \frac{1}{q}) + (\text{geometricterm}) + (\text{atomic phase}) \qquad (1.11)$$

where λ is the laser wavelength; p and η are the gas pressure and ionization fraction, respectively; $\delta n = n_{laser} - n_q$; n_{laser} and n_q are the refractive indices for the laser light and the qth harmonic, respectively, and N_{atm} is the atomic number density at 1 atm. In this expression for the phase mismatch the first (positive) term is related to the medium dispersion and the second (negative) term is due to the plasma dispersion. The geometric term is negative in a waveguide [29] or self guide [30]. For a focused Gaussian beam the geometrical phase shift around the focal point is due to the Gouy phase shift. The sign of the atomic dipole phase term is not fixed since this phase varies with the intensity of the laser field. The phase matching of HHG can be achieved by balancing the various terms presented in equation (1.11) for a zero phase mismatch.

Geometrical Phase Mismatch

This phase mismatch arises primarily from the Gouy phase shift when the laser beam is focused [31]. For a Gaussian beam, the Gouy shift can lead to a phase deviation between the laser wave front passing through the focus and the phase front of a plane wave with the same optical frequency, and is given by

$$\phi_{Gouy}(z) = \tan^{-1}\frac{z}{z_R} \tag{1.12}$$

where $z_R = \frac{\pi w_0^2}{\lambda}$ is the laser Rayleigh length, w_0 is the beam waist radius (for a typical f/50 geometry with 800 nm wavelength, $z_R = \frac{\pi w_0^2}{\lambda} = 2.5\,\text{mm}$) and z is the distance from the beam focus along the axis of propagation.

As a consequence, the phase for the qth harmonic order at each point where it is generated along the z direction is

$$\phi_{Gouy,q} = q\phi_{Gouy} = q\tan^{-1}\frac{z}{z_R} \tag{1.13}$$

The Gouy shift for the harmonics themselves as they propagate along the z direction can be neglected compared to the Gouy shift shown in equation (1.13).

If we define the distance over which the phase given by equation (1.13) shifts by π as the coherence length relating to the Gouy shift, we can estimate this coherence length by taking the derivative of this phase $\frac{d\phi}{dz} = \frac{q}{(z_R + \frac{z^2}{z_R})}$ and setting it equal to π.

Then, this coherence length is given by

$$L_{coh,G}(z) = \frac{\pi(z_R + \frac{z^2}{z_R})}{q} \tag{1.14}$$

Close to the centre of the beam focus (z ~ 0), $L_{coh,G}$ may be rewritten as

$$L_{coh,G}(z) = \frac{\pi z_R}{q} \tag{1.15}$$

For a typical f/50 geometry with 800 nm wavelength, the Gouy coherence length near the centre of the focus can be estimated to be 0.32 mm for the harmonic H25.

According to equation (1.14), $L_{coh,G}(z)$ is larger when z is longer, i.e., the Gouy phase related coherence length is longer at positions away from the focus.

The contribution from the Gouy phase can be controlled by placing the focus at a different position with respect to the gas medium [32]. When the laser is focused before the generating medium, phase matching on the optical axis is efficient leading to spatial and spectral harmonics that are regular. When the focus is at the centre of the medium, the harmonic intensity is low owing to poor phase matching. On the other hand, when the laser is focused after the medium, efficient phase matching is achieved for off-axis emission, resulting in a high conversion efficiency but distorted spatial and temporal profiles.

Neutral Gas Dispersion Phase Mismatch

This phase mismatch contribution is due to material dispersion and is given by (for harmonic q)

$$\Delta k_N = \frac{2\pi q}{\lambda}(n_{laser} - n_{harmonic}) \tag{1.16}$$

We define $\delta n = n_{laser} - n_{harmonic}$ as the difference of the refraction index of the gas per unit atmosphere at the fundamental and harmonic wavelengths. The refractive indices of the noble gases are close to unity and depend on the frequency. For a fundamental wavelength $\lambda = 800$ nm, $n(\lambda) \approx 1 + 2.8 \times 10^{-4}$ and for harmonics q > 21, $n(\lambda_q) \approx (1-10^{-4}) - (1-10^{-6})$ at STP (1 atm and 273K), where λ_q is the wavelength of the qth harmonic.

The refractive index is proportional to the gas pressure (in atm) and when the gas medium is ionized the fraction of ionization is usually given by $\eta = N_e / N_a$, where N_e and N_a are electron and atomic number density respectively. Therefore, equation (1.16) can be rewritten as

$$\Delta k_N = \frac{2\pi q}{\lambda} p\delta n(1-\eta) \tag{1.17}$$

where p is the gas pressure (in atm).

Equation (1.17) is valid for low ionizationlevel $\eta << 1$.

For a gas pressure of 50 Torr, and assuming the ionization fraction in the medium is 1 %, the neutral dispersion phase mismatch for harmonic H25 is found to be around 45 cm^{-1}.

It is worth recalling that when a laser with high intensity propagates through a medium, this also causes a change in the refractive index in proportion to the laser intensity. This change is known as the instantaneous change and given as

$$\Delta n = n_2 I(t) \tag{1.18}$$

where n_2 is the nonlinear refractive index.

Plasma Dispersion Phase Mismatch

When an intense laser beam interacts with a gas medium, not only harmonics but also free electrons are emitted during the ionization process. The free electrons cause plasma dispersion which can reduce the coherencelength, i.e., limit the harmonic yield. The refractive index for a plasma is given by [33]

$$n_P(\omega_0) = \sqrt{1 - \frac{\omega_P^2}{\omega_0{}^2}} \tag{1.19}$$

where ω_0 is the laser frequency, and ω_P is the plasma frequency. The plasma frequency is given by

$$\omega_P = e\sqrt{\frac{N_e}{\varepsilon_0 m_e}} \tag{1.20}$$

where N_e is the density of free electrons, ε_0 is the vacuum permittivity and m_e is the electron mass.

From equation (1.19), it is clear that the refractive index for a plasma is always less than one and depends on the frequency of the laser beam.

From equations (1.19) and (1.20), for a fixed laser wavelength, the refractive index for a plasma is a function of electron density.

Since the laser frequency ω_0 is much larger than the plasma frequency ω_P, i.e., $\omega_P^2 << \omega_0{}^2$, equation (1.19) can be rewritten

$$n_P(\omega_0) \approx 1 - \frac{\omega_P^2}{2\omega_0{}^2} \tag{1.21}$$

If we define $L_{coh,P}$ as the coherence length relating to the plasma dispersion, $L_{coh,P}$ can be calculated from

$$L_{coh,P}(qk_0 - k_q) = \frac{2\pi q}{\lambda_0}[n_P(\omega_0) - n_P(q\omega_0)]L_{coh,P} = \pi$$

then $L_{coh,P} = \dfrac{4\pi^2 m_e}{N_e e^2 \lambda_0 \mu_0 (q - 1/q)}$ (1.22) where $\mu_0 = \frac{1}{c^2 \varepsilon_0}$ is the vacuum permittivity.

If we assume q>>1, then equation (1.22) is given approximately by

$$L_{coh,P} = \frac{4\pi^2 m_e}{N_e e^2 \lambda_0 \mu_0 q} \tag{1.23}$$

From equation (1.23), we can see that the coherence length scales inversely with the harmonic order. In addition, to produce a higher order harmonic, a higher intensity of the fundamental beam needs to be applied. This leads to an increase of the ionization rate and free electron density. Therefore, the plasma phase mismatch becomes more important with increased harmonic order.

Also, equation (1.23) indicates that the coherence length decreases as the free electron density increases. If we assume a linear dependence between the density of the generating medium and the free electron density, this suggests that increasing the gas pressure does not merely lead to a higher harmonic yield. For example, for a gas pressure of 50 Torr, and if we assume that one electron per atom is ionized and there is complete ionization in the medium, the coherence length can be calculated to be approximately 37 μm for harmonic H25 and becomes two times shorter if the gas pressure is increased up to 100 Torr.

From equation (1.22), the phase mismatch due to plasma dispersion is given by

$$\Delta k_P = \frac{\pi}{L_{coh,P}} = \frac{N_e e^2 \lambda_0 \mu_0 (q - 1/q)}{4\pi m_e} \tag{1.24}$$

Moreover, the classical electron radius is given by

$$r_e = \frac{1}{4\pi\varepsilon_0} \frac{e^2}{mc^2} \tag{1.25}$$

and the electron density can be rewritten as

$$N_e = \eta p N_{atm} \tag{1.26}$$

Substituting equations (1.26) and (1.25) into equation (1.24), one can obtain the phase mismatch due to the plasma dispersion

$$\Delta k_P = \eta p N_{atm} r_e \lambda_0 (q - 1/q) \tag{1.27}$$

Using the same conditions which give a coherence length of 37 μm, the plasma phase mismatch of harmonic H25 for complete ionization is about 850 cm^{-1}. This phase mismatch is one hundred times smaller, i.e., 8.5 cm^{-1}, if the ionization fraction in the medium is 1 %.

The contribution from the phase mismatch due to neutral gas dispersion is positive and that due to plasma dispersion is negative. This allows us to reduce the total phase mismatch by varying the laser intensity to adjust the fraction of ionization. However, one must consider the critical value of the ionization fraction, η_{cr}. At this value, the dispersion from the remaining neutral atoms is not enough to balance the dispersion from the free electrons. Hence, to obtain phase matching in high harmonic generation, the ionization fraction must be lower than a critical ionization fraction. From equations (1.17) and (1.27), this critical value can be expressed as

$$\eta_{cr} = \left[1 + \frac{N_{atm} r_e \lambda^2}{2\pi(\delta n)} \right]^{-1} \tag{1.28}$$

We find that the critical ionization is independent of the pressure. In the case of no compensation of the phase mismatch, equation (1.28) allows us to estimate the limiting value of the ionization fraction in the medium in order for phase matching to be achieved. Typical values of η_{cr} for common gases for 800 nm wavelength are quite low, about 0.5% in He, 1% in Ne and 5% in Ar [34]. These critical values limit the phase matched highest order harmonics that can be achieved by simply tuning the pressure: H91, H69, H35 in He, Ne and Ar, respectively. In order to overcome this limitation, alternative approaches need to be used.

Atomic Dipole Phase Mismatch

In HHG, the atomic dipole phase mismatch is an intensity dependent phase mismatch and has been theoretically studied [35] as well as experimentally investigated [23, 32]. Theorigin of this phase is the trajectory acquired by the electron leading to the emission of the qth harmonic in the continuum state. To a first approximation, the phase of the atomic dipole varies linearly with the laser intensity. The laser intensity can vary spatially in both the longitudinal and radial directions. Therefore, the atom dipole phase also varies axially as well as radially. This leads to reduced harmonic emission as well as strong spatial distortion [32]. It may be possible to find situations in which the intrinsic phase can be played against the geometrical phase mismatch in order to improve overall phase matching [32, 35].

For example, in the simple case when the contribution of the Gouy phase shift and dipole phase are considered, the phase mismatch will be minimal when the variation of the harmonic phase with propagation is minimal. When the laser confocal parameter is equal to 5 mm and the focus position is located at z = 0, P. Salières et al. find that when the laser is focused approximately 3 mm before the generating medium, the harmonic phase variation is minimal, i.e., the phase mismatch is a minimum [32].

For a Gaussian laser beam with intensity along the direction of propagation

$$I(z) = \frac{I_0}{\left[1+\left(\frac{2z}{b}\right)^2\right]} \tag{1.29}$$

the contribution of the dipole phase to the phase mismatch on axis of the qth harmonic, $\nabla\varphi_q$, is given by [36, 37]

$$\nabla\varphi_q = -\alpha_q \frac{\partial I}{\partial z} = \frac{8z}{b^2\left[1+\left(\frac{2z}{b}\right)^2\right]^2}\alpha_q I_0 \tag{1.30}$$

where α_q is a coefficient related to the electron trajectories

($\alpha_q \approx 1-5\times10^{-14}\ cm^2/W$ for the short trajectory and $\alpha_q \approx 20-25\times10^{-14}\ cm^2/W$ for the long trajectory) and b is the Rayleigh length.

The dependence of the laser field amplitude and phase on the propagation clearly shows that the dipole phase cannot be established assuming an unperturbed laser beam. In a long interaction length geometry, e.g., a semi-infinite cell, the dipole phase cannot be neglected and the phase gradient of the dipole phase allows the phase mismatch on axis to be reduced in the region where the laser intensity is high. This leads to an enhancement of the HHG efficiency because the atomic response is high. The atomic dipole phase gradient acts as an additional time dependent wave vector which allows a high plasma dispersion to be compensated.

1.1.3.3. Reabsorption of Harmonic Emission

Even if phase matching is achieved, the conversion efficiency is still limited by re-absorption of the harmonic emission in the generating gas medium. This issue plays an important role in loose focusing or in a gas filled hollow-core fiber or a gas cell [38, 39]. For phase matched harmonic emission, the coherent harmonic signal grows as the interaction length and with increasing gas pressure. On the other hand, when a harmonic is travelling through an absorptive medium with no new harmonic generation, the signal is also attenuated in proportional to the negative exponential of these two factors. Thus, when the harmonic yield approaches the maximum value for a particular interaction length and gas pressure, it is not worth trying to improve the output photon flux by increasing the interaction length and the gas pressure since the harmonic signal has reached the re-absorption limit.

E. Constant el al. [38] performed an analysis of the time dependent factors that control the output photon flux including the atomic responses, phase matching conditions and absorption of the medium.

The number of photons emitted on-axis per unit of time and area, N_{out}, is given by [38]

$$N_{out} \propto \frac{4L_{abs}^2}{1+4\pi^2(L_{abs}^2/L_{coh}^2)}\left[1+\exp\left(-\frac{L_{med}}{L_{abs}}\right)-2\cos\left(\frac{\pi L_{med}}{L_{coh}}\right)\exp\left(-\frac{L_{med}}{L_{abs}}\right)\right] \tag{1.31}$$

where L_{med} is the interaction length, $L_c = \pi/\Delta k$ is the coherence length, and L_{abs} is the absorption length (strongly dependent on the wavelength, i.e., less absorption for the higher order harmonics).

Studying the evolution of N_{out} as a function of the medium length for several coherence lengths, they found that the overall optimizing conditions in the absorbing medium are

$$L_{med} > 3L_{abs} \text{ and } L_c > 5L_{abs}$$

In the case of no absorption, the harmonic intensity grows quadratically with the medium length as indicated by equation (1.10).

These optimizing conditions are time dependent and strongly influenced by ionization.

1.1.3.4. Phase Matching in High Harmonic Generation

Since phase matching is a critical issue of high harmonic generation, so far a lot of work has been done to improve the phase matching, leading to enhancement of the harmonic conversion efficiency. In principle, phase matching in HHG can be obtained by balancing the different phase mismatch terms shown in equation (1.11) to give a zero total phase mismatch.

We recall that in equation (1.11) the neutral dispersion phase mismatch is positive, the plasma phase mismatch is negative and the geometrical phase mismatch is positive for a focused Gaussian beam and negative for a waveguide or a self guide. Thus, these terms can be balanced by simply adjusting the ionization fraction through the fundamental laser energy or adjusting the gas pressure. Indeed, this traditional phase matching technique is only applicable for a low ionization fraction which is less than the critical level. In order to generate the higher order harmonics or improve the conversion efficiency, higher laser intensities need to be applied. However, at these intensities, the ionization fraction is higher than the critical value; then the dispersion of the free electrons cannot be compensated by simply controlling the pressure. Therefore, some remarkable approaches relying on phase matching issues to improve the conversion are needed.

One very effective approach for obtaining true phase matching in HHG was developed by H. Kapteyn et al. [17, 29]. In their work, they investigated the phase matched harmonic generation in a hollow wave guide with a 800 nm fundamental laser pulse. They could obtain a conversion efficiency up to 10^{-6} to 10^{-5}, an increase of 100 times over previously measured efficiencies due to the phase matching. By controlling several experimental parameters, such as the gas pressure, the gas species and the size and spatial mode of the waveguide, phase matching could be achieved as long as the total percentage of the ionized atoms is still below the critical level. For a waveguide, three major phase mismatch terms are two pressure dependent terms (neutral dispersion and plasma dispersion) and a pressure independent term (geometrical). The phase correction of the waveguide has the same sign as the plasma dispersion phase mismatch and hence the regime to achieve phase matching corresponds to a balance between the dispersion of the neutral atoms and that of the waveguide and plasma. Another advantage of this configuration is that the geometrical phase compensation term remains unchanged with propagation distance leading to the fact that phase matching could be achieved over a long propagation distance.

J. F. Hergott et al. [40] studied high harmonic generation at a high laser pump energy using a long focal length lens (f = 2 m or 5 m) with a gas jet. In this configuration using higher laser energies and looser focusing, the phase mismatch terms were cancelled out by controlling the laser aperture and the length, pressure and position of the gas medium resulting in an increase of at least one order of magnitude of the harmonic energies (up to microjoule harmonic energies in xenon) compared with previously reported values. Despite the fact that a higher pump laser energy was applied, the conversion efficiency was also enhanced by a factor of 6 in their experiment.

Y. Tamaki et al. [41] have demonstrated phase matched high harmonic generation by a self guided beam in a gas cell filled with neon at low pressure (< 50 Torr). Based on this method, they could increase the interaction length up to 7 mm and thus the harmonics in the cutoff region could be magnified by 2 orders of magnitude compared to those in the plateau.Moreover, the conversion efficiencies in the cutoff region were enhanced up to 10^{-6}. In this scheme, since the atomic dipole phase was constant over the entire propagation length, the harmonic spectrum exhibited a good contrast and was smooth.

C. Altucci et al. [42] experimentally investigated the use of Bessel Gauss beams to generate the harmonics in gas jets. By matching the gas medium length to the range of the Bessel Gauss beams, it was shown that the conversion efficiencies for low order harmonic generation were higher than could be achieved with Gaussian beams. Use of a Bessel Gauss beam for high order high harmonic generation in a gas cell was also performed in the present dissertation.

L. V. Dao et al. [43, 44] reported high harmonic generation by using a freely propagating fundamental beam in a gas cell which provides a longer interaction length to obtain the same benefits of homogeneous phase matching as in hollow core fibres. In addition, they could achieve a high photon flux (up to 10^{10}-10^{12}photons/cm^2/s), high spatial coherence (up to 0.95) and a good spatial beam profile.

In summary, there have been many techniques to overcome the phase mismatch problems for improving the efficiency of converting laser light into high harmonics. However, each of them has both advantages and disadvantages. For example, a hollow waveguide is limited to low gas pressures. Also, coupling to the waveguide limits the maximum pump energy and energy losses are unavoidable. A long focal length lens and a semi-infinite gas cell which provides long interaction geometry faces challenges of controlling the phase matching process because of plasma dispersion and neutral dispersion. Also, defocusing effects can lead to severe distortion of the harmonic emission. Thus, phase-matching techniques for HHG are still an open and debated topic.

1.1.4. Quasi Phase Matching in High Harmonic Generation

Usually, the higher fundamental laser intensity required to generate higher order harmonics creates a much higher ionization fraction in the medium. Thus, an uncompensated phase mismatch caused by plasma dispersion prevents the efficient frequency conversion process. When true phase matching cannot be achieved, an alternative approach known as quasi phase matching (QPM) has been recently implemented to significantly extend the range of photon energies over which harmonic production is efficiently generated. Instead of trying to obtain the true phase matching condition, in the QPM technique the harmonic emission from those regions having destructive phase is eliminated leading to enhancement of the harmonic intensity.

Quasi Phase Matching was first proposed by Armstrong et al. [45]. In their work, the phase mismatch is periodically corrected by introducing a periodicity corresponding to twice the coherence length in the nonlinearity of the medium. Up to now there have been several approaches for Quasi Phase Matching of the HHG process, either to modulate the driving laser field or the generated harmonic field.

A modulated waveguide can be used to create a periodic change in the fundamental laser intensity [16]. The key to this scheme is the additional term ($K_m = 2\pi q / \Lambda$, where q is the harmonic order and Λ is the modulation period) resulting from the modulation period to the total phase mismatch of the HHG process (Δk). This technique makes it possible to use neon gas as the nonlinear medium to generate harmonics in the water window region [16]. Therefore, the harmonic signal would be enhanced when $K_m \approx \Delta k$. A higher value of K_m, i.e., a shorter modulation period (Λ), can compensate for a larger phase mismatch created by a high level of ionization. However, due to the fixed modulation period for a given fibre and propagation effects of the waveguide such as mode beating and group velocity dispersion, the range of harmonics is limited, i.e., there is only partial compensation for the phase mismatch.

A quasi phase matching technique studied by J. Seres et al. was achieved by periodic modulation of the gas pressure in a free space focus geometry [46]. The principle of this configuration is shown in figure 1.5.

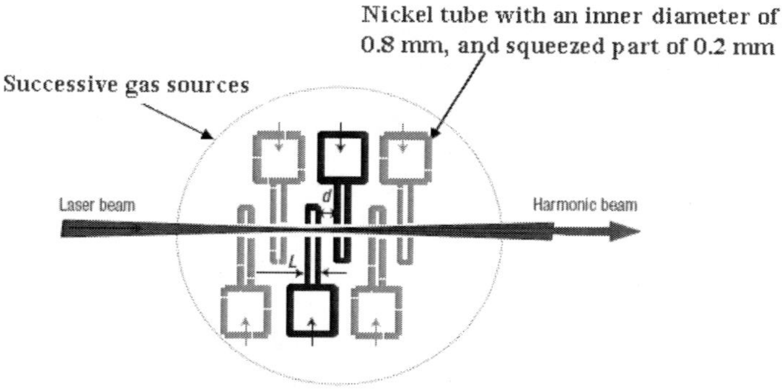

Figure 1.5.Successive gas sources for QPM. dis the distance between the two sources and can be varied. The interaction between the laser beam and target atomstakes place within a couple of nickel tubes.

It is known that the intensity of the qth harmonic (I_q) is linearly proportional to the number atom density (N_a). However, further increases of N_a, i.e., higher gas pressure, would reduce I_q since the phase mismatch causes destructive interference of the harmonic production in the first half and second half of the source. In the theoretical background of this approach, by subdividing the overall interaction length L into M sections of thickness of L/M, and separating them so that diffraction induced changes in the laser phase and amplitude between two adjacent sections can shift the atomic dipole phase by π, the presence of destructive interference can be shifted to higher atomic densities. Therefore, the possibility of this technique is to improve the harmonic yield by increasing the atomic density by a factor of M. In the experimental work, by optimizing the distance between two regions of high gas density along the laser propagation, a significant enhancement of the harmonic yield, especially at very high photon energy (~300 eV), has been achieved. However, this method still has some limitations that need to be considered like defocusing of the laser beam by the radial variation of the free electron density, and some experimental challenges.

Another quasi phase matching technique involves the use of counter-propagating laser pulses. Instead of trying to modulate spatially either the laser intensity or the atom density, in this approach the phase of the driving pulse can be controlled optically. The counter-propagating pulse can be used to intersect the driving laser pulse and the timing and duration of the counter-propagating pulse can be chosen to suppress the harmonic emission from out-of-phase regions in the medium [47, 48, 49] (figure 1.6). Then, the remaining regions constructively interfere and the harmonic production is strongly enhanced.

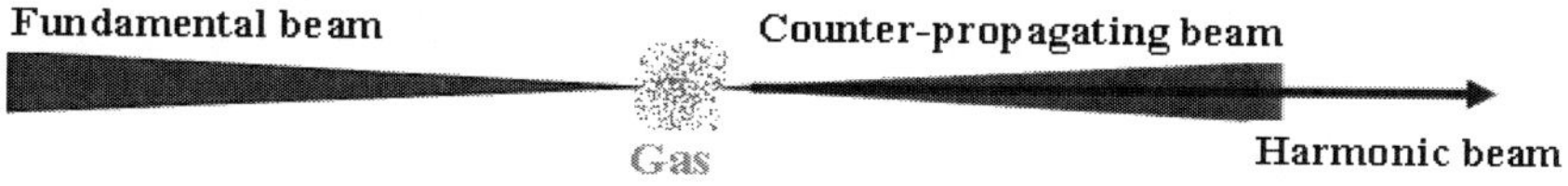

Figure 1.6.Quasi Phase Matching by using a counter-propagating beam

In a theoretical analysis of the influence of a relatively weak counter-propagating beam J. Peatross et al. [47] showed that in the interaction region between the fundamental pulse and the counter-propagating pulse there is a standing wave modulation in both the intensity and phase of the generating laser field with a period of half the fundamental laser wavelength. Since the harmonic process is extremely nonlinear, a tiny spatial variation in the phase of the generating laser field can lead to a significant spatial phase variation of each of the harmonics, no matter whether the intensity of the counter-propagating beam is very small compared with the fundamental beam. This phase modulation in turn is able to cause significant phase mismatches over a distance of half a laser wavelength for the harmonics. As a result, the harmonic production is turned off. In other words, these rapid modulations can microscopically eliminate coherent build up of the harmonic emission. Based on this analysis, it is clear that the harmonic generation can also be enhanced by using the counter-propagating beam to suppress emission from selected regions having destructive phase interference. Moreover, due to the very weak intensity, the counter-propagating beam does not change the properties of the medium, and thus it is possible to apply this approach accompanied with other phase matching techniques.

In experimental work, S. L. Voronov et al. [48] have demonstrated strong disruption of the phase matched harmonic generation by a relatively weak counter-propagating beam. On the other hand, under poor phase matching conditions, they showed that significant enhancement of the harmonic yield was achieved due to the presence of the counter-propagating beam. However, they only investigated harmonic emission regions where conventional phase matching techniques have already been achievable. J. B Madsen et al. [49] investigated high harmonic generation in the presence of a near counter-propagating beam and a perpendicular beam. It was also shown that the relatively weak interfering light can substantially turn off the harmonic emission due to phase modulation of the generating laser field. In addition, they achieved enhancement of the harmonic yield in the presence of the interfering light under poor phase matching conditions. The advantage of the interfering light method compared with the counter-propagating beam method is the prevention of potentially destructive feedback to the laser amplifier system. More recently, X. Zhang et al., [50] demonstrated the enhancement of HHG in a hollow waveguide using a three pulse sequence as the counter-propagating pulse train. In this work, the intensity of the harmonic emission around 70 eV in argon gas was improved by up to two orders of magnitude compared to the conventionally phase matched emission from helium gas.

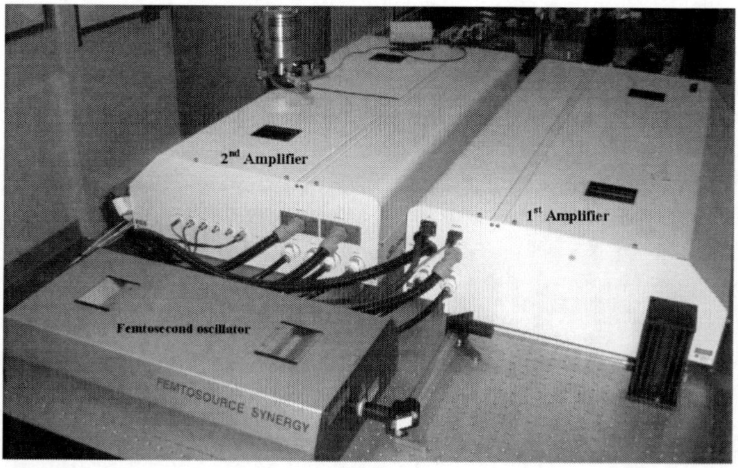

Figure 1.7.High power femtosecond laser system.

1.2. GENERAL EXPERIMENTAL SETUP

1.2.1. High Power Femtosecond Laser System

The high power femtosecond laser used in our experiments is a chirped pulse amplification system. The system consists of a femtosecond oscillator and two multi-pass power amplifiers as shown in figure 1.7.

1.2.1.1. Femtosecond Oscillator

A mirror dispersion controlled titanium:sapphire (Ti:Sa) oscillator (Femtolasers Femtosource Synergy-20) has been chosen for generation of the femtosecond pulses. The mirror dispersion control set comprises a chirped mirror for broadband intra cavity group delay dispersion in the Ti:Sa oscillator and low dispersion quarter wave mirrors for coupling the pump beam into the resonator and the mode locked pulses out of the resonator. A combination of the mirror dispersion control set and a thin highly doped Ti:Sa crystal allows the generation of optical pulses with high quality, stability and reproducibility. The heat from Ti:Sa crystal can be removed by water cooling. The Ti:Sa crystal is pumped by a green pump laser (Coherent Verdi-5). The Verdi-5 laser system is a compact solid state diode-pumped, frequency- doubled Nd:Vanadate (Nd:YVO_4) laser that provides the fundamental mode of a TEM_{00} mode and a pump power of 4-5 W at a single frequency of 532 nm. The output beam of the pump laser is vertically polarized and then rotated horizontally for the seed laser by using a periscope. The periscope also allows the beam height to be adjusted.

The output from the oscillator has anenergyof 8.4 nJ per pulse. The specifications of the laser output after the oscillator shown in Table 1.1 are given by the manufacturer.

Table 1.1. Specifications of the laser output of the seeding laser

Pulse duration	<20 fs
Spectral width @ 798 nm	76 nm
Output power (average)	630 mW
Output energy @ 75 MHz	8.4 nJ

1.2.1.2. Amplification System

The output of the femtosecond oscillator is passed through two multipass Ti:Sa amplifier systems (Quantronix Odin-II HE amplifier and Cryo Add-on amplifier)to produce output pulse energies up to ~ 10 mJ.

Quantronix Odin-II HEAmplifier

The first amplifier, i.e., Odin-II HE, allows amplification of the pulse energy of sub nJ energy generated by an oscillator to a level of a few mJ. The typical operating repetition rate is 1 kHz. The Odin-II HE comprises five main components: pump laser (Quantronix Darwin-527), pulse stretcher, 8-pass amplifier, power amplifier and pulse compressor. Both the grating stretcher and the compressor are arranged in off-Littrow geometry. The technique of chirped pulse amplification is based on stretching the short pulse in time by introducing a frequency chirp onto the pulse which increases the duration by a factor of 10^3-10^4, amplification of the broadened pulse and then compression of the pulse. The mechanism of this technique follows in steps: Firstly, the pulse is stretched by multi-pass propagation of the pulse into a single grating which exhibits negative dispersion. The stretching of the pulse leads to a significant reduction of its intensity which removes the tendency of gain medium damage due to the high intensity of the short pulse. Secondly, the beam passes through the gain medium multiple times without the use of a cavity. Compared to a regenerative amplifier, amplified spontaneous emission can be suppressed to a greater degree since the optical path is not a resonator. Therefore, typically, this results in a higher gain per pass. In the Odin-II HE system, the optical pulse is initially amplified in an 8-pass amplifier where a Pockels cell is installed to select a single pulse from the oscillator pulse train with a 1 kHz repetition rate after the first 4-passes, and then the additional 2-pass amplifier is used as a power amplifier. Both amplification stages are pumped by the same pump laser by dividing the energy between the two states with a beam splitter. The pump laser is a diode pumped, Q-switched frequency doubled neodymium:yttrium lithium fluoride laser. Finally, the amplified pulse is compressed by using a grating pair which exhibits positive dispersion for a direct output with ~ 30 fs duration and ~ 2.6 mJ energy as specified by the manufacturer. If a higher energy pulse is required, the amplified pulse is passed directly to the second amplifier instead of compression.

Quantronix Cryo Add-on Amplifier

The second amplifier, i.e., Cryo Add-on, is usually used after a front-end amplifier like the Odin-II HE and can amplify the output of the pulse energy up to 10 mJ by using a cryogenically cooled crystal.

It is known that the thermal load in the Ti:Sa crystal has a similar distribution as the pump laser profile. This causes a temperature gradient in the crystal which induces a refractive index distribution called a thermal lens. The thermal lens leads to low efficiency and a poor beam profile of the amplified laser beam. In an amplifier with a moderate power output, the cavity length is much less than the focal length of the thermal lens, and therefore the beam is less affected. However, in a high power amplifier in which the cavity length is comparable to the focal length of the thermal lens, the effect of the thermal lens becomes critical and needs to be minimized. Thus, in the Quantronix Cryo-cell, the crystal is cooled below 100 K. The Cryo Add-on operates at 1 kHz repetition rate and comprises five main components: two pump lasers, two-pass amplifier, Cryo-crystal cell and pulse compressor. The cryogenically cooled crystal must be kept in a vacuum to avoid icing on the crystal. Thus, its structure consists of a laser crystal attached to a cold finger and a vacuum chamber containing all the cold components. A compact ion pump is used to maintain a vacuum in the chamber.

In the second amplifier, the amplified pulse without compression from the first amplifier is passed to the two-pass amplifier and then is compressed for a very high output power ~ 10 mJ. The specifications of the laser output after the Cryo Add-on amplifier are given in table 1.2.

Table 1.2. Specifications of the laser output after the second amplification stage

Centre wavelength	805 nm
Pulse energy	10 mJ
Pulse width ($sech^2$)	30 fs
Beam size	12 mm
Beam quality (M^2)	1.6
Instability	< 1%
Repetition rate	1 kHz
Output polarization	Horizontal

1.2.2. Fundamental Experimental High Harmonic Generation Setup

1.2.2.1. Fundamental Arrangement and HHG Process

The fundamental experimental arrangement for generation and detection of the harmonic radiation is shown below.

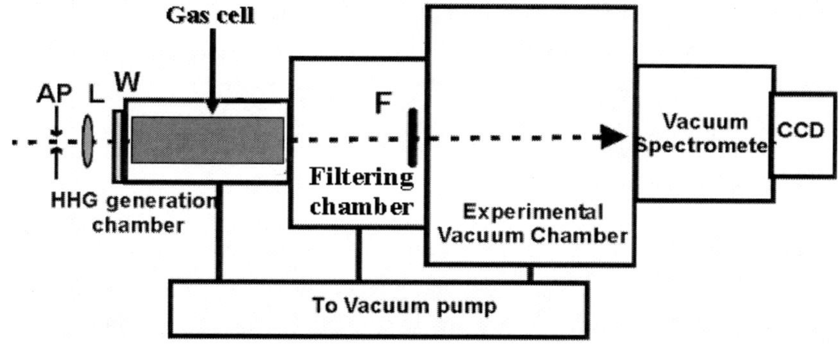

Figure 1.8.Schematic diagram of the fundamental experimental setup for generation and detection of the harmonic radiation.

The output of the Cryo Add-on amplifier, known as the fundamental laser beam, is passed directly to the high harmonic generation apparatus. This beam is focused by a focusing lens into a 150 mm long gas cell (semi-infinite gas cell) with a glass window at the entrance and an ~ 0.1 mm pinhole at the exit which is used to separate the vacuum chamber from the gas filled cell. The effective peak intensity at the focus is approximately 10^{14}-10^{15} W/cm^2. The aperture which is placed before the focusing lens is used to control the fundamental laser beam. Changing the laser beam by adjusting the aperture affects not only the intensity but also the beam profile of the harmonics. The gases that can be used are argon, neon, helium, or molecular gases, like nitrogen, oxygen.

The pressure, aperture diameter, energy and chirp of the laser pulses, and the position of the laser focus relative to the exit of the gas cell are optimized in an iterative procedure for maximum flux of all available harmonics. The pressure in the vacuum chamber is kept less than 10^{-3}-10^{-4} Torr for the experimental chamber. The high harmonics generated on-axis or close to the optical axis pass through a filter (aluminum, zirconium), which removes the fundamental beam. For measurement of the HHG spectrum, a grazingincidence spectrometer (GIMS #4 – set point) with a grating (150, 300, 600, or 1200 grooves/mm) is inserted into the beam path. A 13 mm x 13 mm CCD chip with 13 μm pixel size (Princeton Instruments PIXIS-XO) is used as the detector.

1.2.2.2. Fundamental HHG Apparatus

Semi-infinite Gas Cell

The semi-infinite gas cell is defined by a ~ 2 mm thick fused silica glass window as the entry point for the laser pulses and a ~1 mm thick aluminium foil with a ~ 100 μm pinhole as the exit point of the harmonic signal. The input window is placed close to the lens to avoid continuum generation and the output pinhole on the aluminium foil is drilled online by the laser to ensure the alignment of the laser beam. The length of the semi-infinite gas cell is chosen according to the focusing lens used for the experiment and normally this length is of order ~ 10 cm, for example, the length is ~ 15 cm for a 300 mm focusing lens. Thus, the length of the gas cell is much longer than the interaction length of the laser field in the nonlinear medium which is usually less than 1.5 cm in our experiments. The focus of the lens can be controlled manually by means of a micrometer screw or electronically by means of a linear stepper motor (T-LA60A Zaber stepper motor). The pressure inside the gas cell which is measured by a pressure gauge is controlled by a combination of a mass flow controller and a micrometer screw handled pressure valve. The immediate space around the exit of the gas cell is pumped by a piston pump at a pressure between 10^{-2} and 10^{-1} Torr and separated from the filtering chamber by a vacuum gate (figure 1.12).

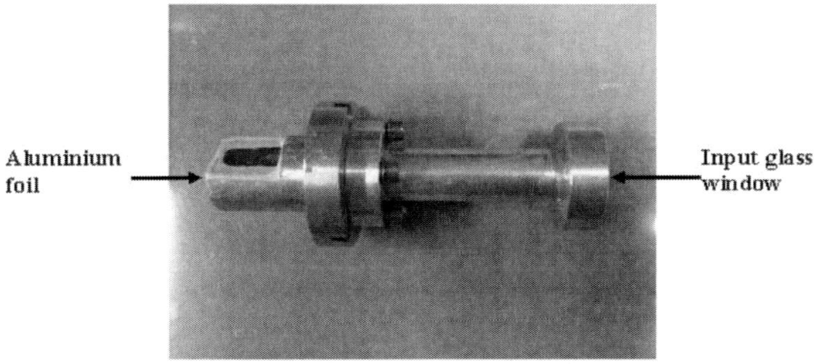

Figure 1.9. Semi-infinite gas cell.

Filtering Chamber

In this chamber (figure 1.12), the harmonic emission is passed through two spatial filter sets and a spectral filter. Firstly, the combination of the fundamental beam and harmonic beam reaches the first set of spatial filters which have several pinholes of diameters 100 μm to 5 mm to reduce thermal impact on the ultra-thin metal foil. A linear motion feed-through mounted on a flexible bellow on the top of the filtering chamber in combination with a spring loaded micrometer screw allows us to adjust the spectral filters easily.

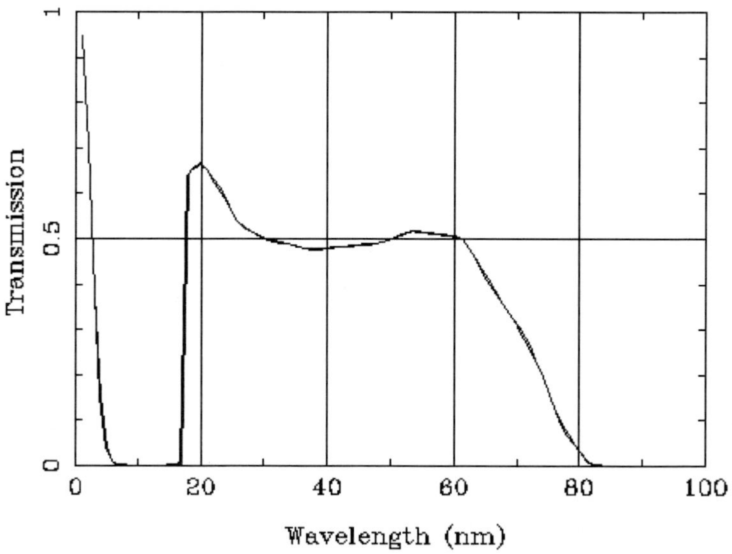

Figure 1.10. Transmission spectra of the Al foil with a thickness of 300 nm (calculated at http://henke.lbl.gov/optical_constants/filter2.html).

After passing the first spectral filter set, the fundamental beam is isolated from the harmonic beam by means of the spectral filters which have a high transmission for the particular harmonic emission but are opaque for the fundamental wavelength (~805nm). The spectral filters are mounted on a filter wheel which is controlled by an electric motor. Depending on which XUV range we would like to observe, different kinds of ultra-thin foils having thicknesses of 100 nm to 300 nm are employed. For example, an aluminium foil is used for harmonic generation with argon gas and a silver foil or a zirconium foil is used for harmonic generation with helium gas.

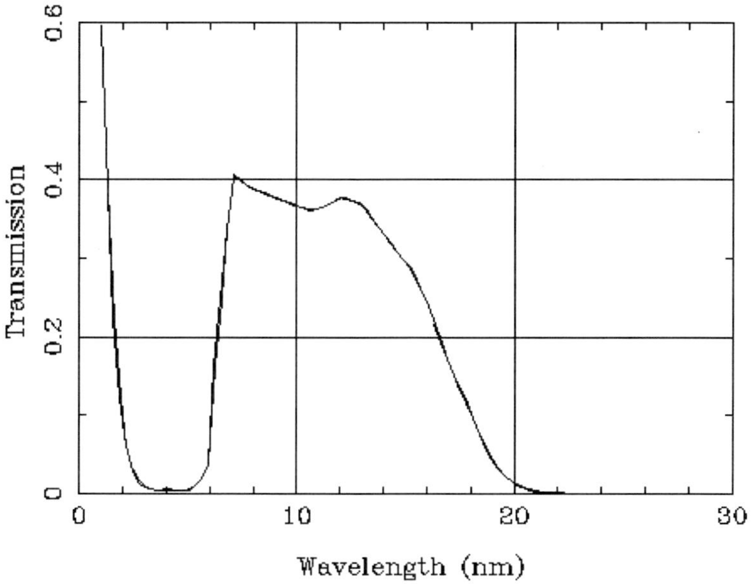

Figure 1.11. Transmission spectra of the Zr foil with a thickness of 300 nm (calculated at http://henke.lbl.gov/optical_constants/filter2.html).

After isolation from the fundamental beam, the harmonic beam passes through the second spatial filter set which provides a precise spatial filtering for the harmonic signal. This filter set consists of several pinholes of precise diameters 500 μm to 50 μm and is mounted on a linear motion feed-through for the y direction. Precision control in the x direction is implemented using a piezoelectrically driven motor. The filtering chamber is pumped by a turbo molecular drag pump (Pfeiffer Vacuum TMU 262) at pressures between 10^{-3} to 10^{-4} Torr which is measured by a pressure gauge.

Experimental Chamber

After passing through the filtering chamber, the harmonic signal passes to the experimental chamber which is connected to the filtering chamber by a 2 cm diameter hole (figure 1.12). In the experimental chamber, all apparatus, like the focusing mirror and the sample holder, used for the coherent diffractive imaging (CDI) experiment (presented in chapter 5) can be installed.

The sample holder is mounted on the linear stepper motor (T-LA60A Zaber stepper motor) for controlling motion in the z direction. This stepper motor is mounted on a set of x/y-translation stages which is controlled by two other linear stepper motors. Thus, the sample holder can be monitored precisely at any point across the spatial profile of the harmonic beam. All motors are controlled from outside of the experimental chamber by using a LabView based software module. A focusing mirror holder is mounted on a post holder which is fixed to another mirror holder. This mirror holder is mounted on two piezoelectrically driven motors for controlling the angle of the focusing mirror. The set consisting of the mirror holder and these two motors is fixed with a linear motion feed-through bellow on top of the experimental chamber to control the height of the mirror. The experimental chamber is also pumped by a turbo molecular drag pump (Pfeiffer Vacuum TMU 262) at a pressure between 10^{-4} Torr and 10^{-5} Torr.

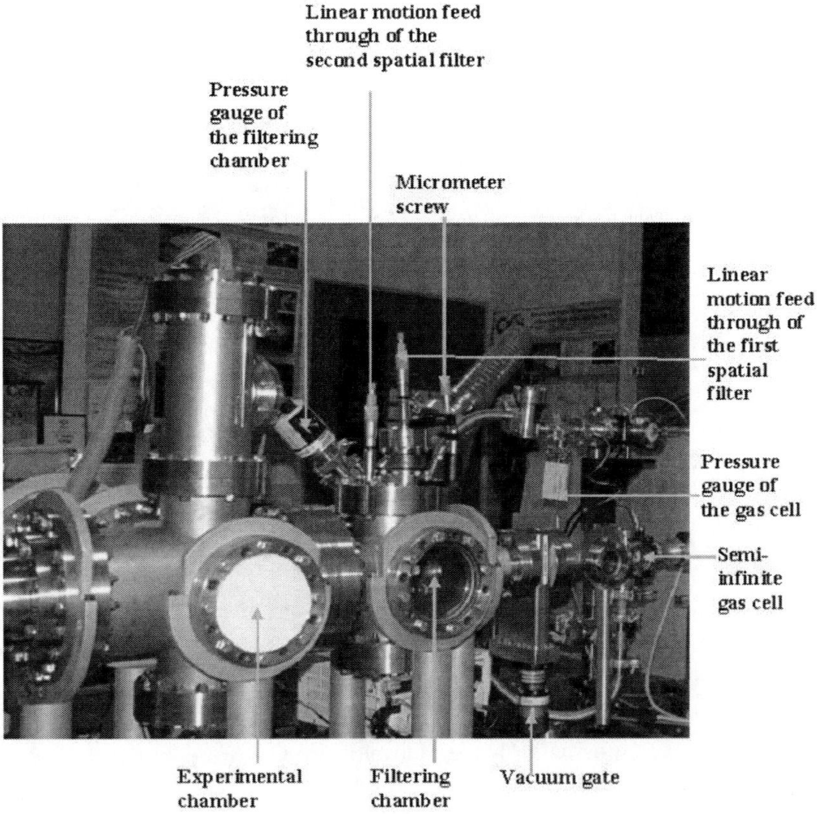

Figure 1.12.Fundamental HHG experimental apparatus.

Extreme Ultraviolet Spectrometer

A extreme ultraviolet spectrometer (Setpoint GIMS#4) with a grazing incidence diffraction grating can be inserted into the harmonic beam path between the experimental chamber and the CCD detector to capture the harmonic spectra (figure 1.13). The spectrometer has a micrometer driven entrance slit. In our design, the expected broad harmonic spectral range can be acquired by precise choice of the grating position. Rotation of the grating is implemented manually by means of a micrometer screw. With four gratings, the spectral range covers from 4.4 nm to 85 nm. However, it is possible to extend the spectral range on both sides with losses in spectral resolution. Based on the spectral range which needs to be investigated in our experiments, a suitable grating is selected. Gratings with a small width of the grooved part are needed for sufficient spectral resolution.

For the spectral range (λ) 4.4 nm to 85 nm, angles φ (incident angle) and ψ (diffraction angle) are varied as $\varphi = 85^0 \pm 1^0$ and $\psi = 81^0 \pm 1^0$. The blaze angle is equal to $\theta = (\varphi - \psi)/2 = 2^0$ for all three gratings. The angle of deviation is equal to $\alpha = 166^0$.

The angular dispersion D_a is given by $D_a = -2d\cos(\alpha/2)\cos(\alpha/2 - \psi)$, where d is the grating period. The linear dispersion D_b is given by $D_b = d\cos\psi / r_b$, where r_b is distance between the grating and the exit slit (focal plane).

Table 1.3. Parameters of the grating 300 grooves/mm

φ (⁰)	ψ (⁰)	λ (nm)	D_a ($\mathring{A}$/rad)	D_a ($\mathring{A}$/mm)	D_b ($\mathring{A}$/mm)
84	82	14.18	8123.1	79.96	29.2
85	81	28.36	8119.2	79.92	32.8
86	80	42.52	8113.2	79.84	36.4

The values of φ, ψ, λ, D_a (angular dispersion) and D_b (linear dispersion in the focal plane) of the grating 300 grooves/mm is presented in table 1.3

Details of four gratings are listed below where the wavelength range is Δλ, the energy range is ΔE, and the spectral resolution is δλ.

Table 1.4. Parameters of the four gratings

Grating (grooves /mm)	Δλ (nm)	ΔE (eV)	Δλ (Å)
1200	4.4 – 10.6	281 – 117	0.5
600	8.9 – 21.3	139 – 58	1.0
300	17.7 – 42.5	70 – 29	2.0
150	35.4 – 85.0	35 – 15	4.0

Figure 1.13.CCD camera and spectrometer.

Detector

A charged coupled device camera system 1024 × 1024 imaging array of 13 μm pixel size (Princeton Instruments PIXIS-XO) is used as the detection device for the high harmonic emission (figure 1.13). The high speed readout is 2 MHz (16 bit) for rapid image acquisition and the low speed readout is 100 kHz (16 bit) for high signal to noise ratio. For normal operation, the CCD is thermoelectrically cooled to a temperature of -30^0 C to allow a low dark current. The CCD has a USB 2.0 interface and the acquisition of the data is controlled by Princeton Instruments WinView software. The data processing is performed by software modules written by LabView programming. The HHG photon number N_{photon} is calculated from

$$N_{photon} = \frac{c\eta}{QE} \times \frac{3.65}{E_{ph}} \tag{1.32}$$

where c is the number of counts per CCD pixel, η depends on the setting of the hardware gain given by the manufacturer, QE is the quantum efficiency of the detector which is nearly constant over the range of photon energies 20–200 eV, and E_{ph} is the photon energy.

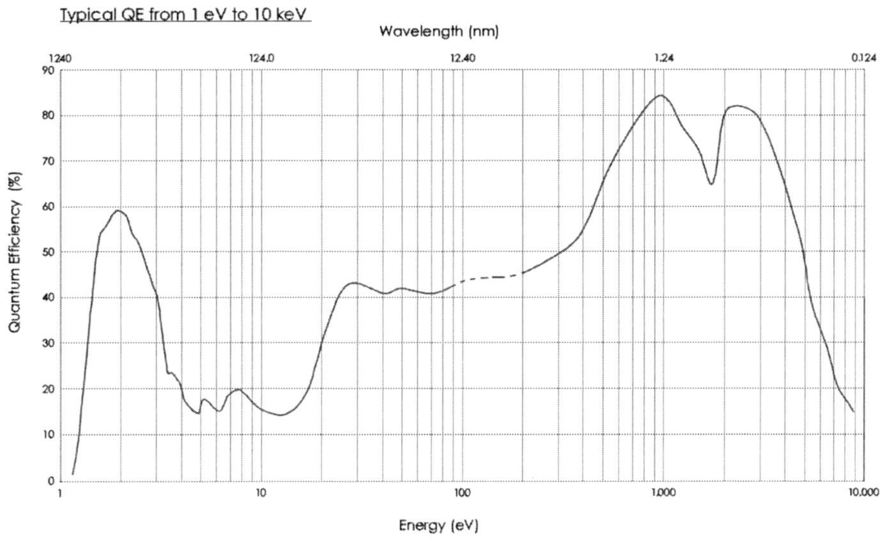

Figure 1.14.Quantum efficiency of the camera used for the experimental work(Princeton Instruments PIXIS-XO).

CONCLUSION

In this chapter, we have shown an overview of history and principle physical processes of the high harmonic generation technique. The influence of macroscopic parameters on harmonic emission and the previous studies on enhancement of phase matching and quasi phase matching in HHG have also been studied. In addition, we have shown required equipments for high harmonic generation and described the general experimental arrangements for generation and detection of the harmonic radiation in a semi-infinite gas cell.Based on the phase matching effects of the nonlinear conversion process, in this book we will address the working on developing new ways to manage phase matching conditions in order to improve the quality of the HHG source including the conversion efficiency, the coherence and the cut-off photon energy.

Chapter 2

ENHANCEMENT OF HIGH HARMONIC GENERATION SOURCE

2.1. ENHANCEMENT OF HIGH HARMONIC GENERATION SOURCE BY USING A COMBINATION OF A LENS AND AN AXICON

2.1.1. Introduction

As mentioned in section 1.1.3, the phase matching between the fundamental laser field and the harmonic field plays a very important role for HHG. In general, the phase mismatch along the optical axis is given by equation (1.11). Many techniques have been proposed to deal with this problem.

Recently, the use of a non-Gaussian or a modified Gaussian beam for generation of high-order harmonics has been studied with a lot of interest [27, 52, 53, 54]. Nonlinear processes at the waist of a Gaussian beam occur in a collinear excitation geometry whereas in the case of a Bessel beam they take place in a non-collinear geometry. A Bessel beam that is generated by an axicon or a suitable hologram can be considered as a superposition of an infinite number of equally weighted plane waves whose wave vector lies on a conical surface centred on the propagation axis. In theoretical studies, a combination of Bessel and Gaussian beams can be used to enhance the conversion efficiency in HHG by adjusting the geometric wave vector mismatch resulting in the elimination of the total phase mismatch [53]. In experimental investigations, Bessel-Gaussian (BG) beams have also been used for enhancement of phase matching in the generation of very low harmonic orders [42].

In the first part of this chapter, we show that by using a combination of lens and axicon to change the geometry of the fundamental beam leading to the improvement of the phase matching, an enhancement of the phase matched generation of harmonics in the cut-off region with higher photon flux and better spatial beam profile can be achieved.

2.1.2. Phase Matching Condition in a Bessel Gauss Geometry

2.1.2.1. Characteristics of a Bessel Gauss Beam

Here, we demonstrate the characteristics of the Bessel Gaussian beam which can be produced by sending a Gaussian laser beam through an axicon placed in the waist plane of the Gaussian beam.

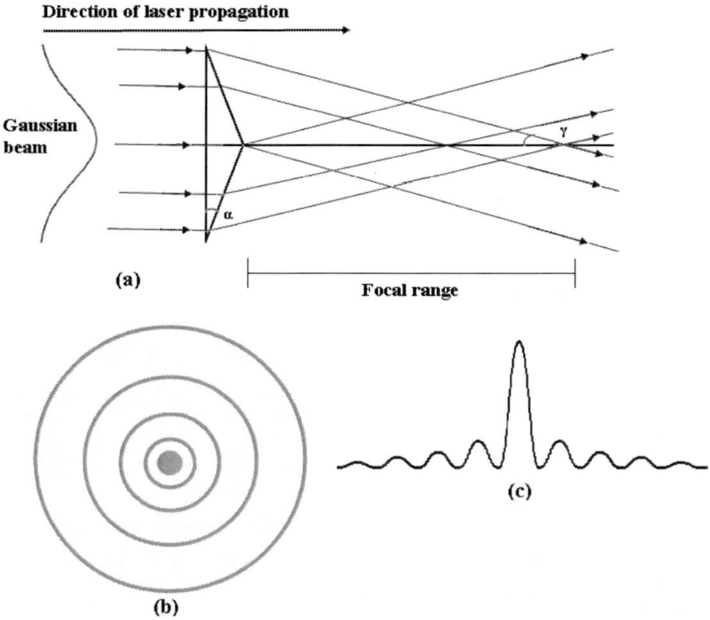

Figure 2.1.(a) A Bessel Gaussian beam is produced by sending a Gaussian beam through an axicon placed in the waist plane with base angle of α and the phase front tilts at an angle of γ; (b) The cross section of the Bessel beam in the x-y plane in the focal area; (c) The radial intensity distribution of the Bessel beam in y-z plane in the focal area.

The spatial distribution of the electric field for a Bessel Gaussian beam can be given as a function of the radial and longitudinal cylindrical coordinates, r and z, and is written below [52, 53].

$$E_{BG}(r,z)=\frac{E_f}{\sqrt{1+\left(\frac{2z}{b_{BG}}\right)^2}}J_0\left(\frac{kr\sin\gamma}{1+i\frac{2z}{b_{BG}}}\right)\exp\left[-\left(\frac{k(r^2+z^2\sin^2\gamma)/b_{BG}}{1+(2z/b_{BG})^2}\right)\right]$$
$$\times\exp\left[i\left(kz\cos\gamma-\tan^{-1}\frac{2z}{b_{BG}}\right)\right]$$
$$\times\exp\left[i\frac{2z}{b_{BG}}\left(\frac{k(r^2+z^2\sin^2\gamma)/b_{BG}}{1+(2z/b_{BG})^2}\right)\right]$$

(2.1)

where $E_f = E_{BG}(0,0)$ is the field amplitude on-axis ($r = 0$) in the waist plane $z = 0$, J_0 is the Bessel function of zero order, b_{BG} is the confocal parameter, k is the wave number, and γ is the focusing half angle, i.e., the exit angle of the Bessel Gaussian beam.

To maintain E_{BG} as a polarized beam, the angle γ must be restricted to be small enough. The radius of the central spot of the BG beam is given by

$$r_0=\frac{2.405}{k\sin\gamma}$$

(2.2)

When $\gamma=0$, i.e., without the axicon, equation (2.1) takes on a simple form and expresses the field of a Gaussian beam with a confocal parameter $b_G \equiv b_{BG}$

$$E_G(r,z)=\frac{E_f}{\sqrt{1+\left(\frac{2z}{b_G}\right)^2}}\exp-\left(\frac{k(r^2/b_G}{1+(2z/b_G)^2}\right)\exp\left[i\left(kz-\tan^{-1}\frac{2z}{b_{BG}}+\frac{2z}{b_G}\left(\frac{k(r^2/b_G}{1+(2z/b_{BG})^2}\right)\right)\right]$$

(2.3)

When $b_{BG} \to \infty$, E_{BG} in equation (2.1) reduces to an ideal Bessel beam which is expressed by

$$E_B(r,z) = E_f J_0(kr\sin\gamma)\exp(ikz\cos\gamma) \tag{2.4}$$

Equation (2.4) has the form of a diffractionless beam, where J_0 is the modified Bessel function of zero order.

The relation between these confocal parameters b_{BG} and b_G is given by [53]

$$b_G = b_{BG}\exp\left(-\frac{b_{BG}\pi\sin^2\gamma}{2\lambda}\right)J_0\left(\frac{b_{BG}\pi\sin^2\lambda}{2\lambda}\right) \tag{2.5}$$

We note that b_G in equation (2.5) is the confocal parameter of the Gaussian beam having the same intensity E_f and the same pulse duration as the Bessel Gaussian beam. It has been shown that b_{BG} is much larger than b_G, i.e., the Rayleigh range of the Bessel Gaussian beam is much longer than that of the associated Gaussian beam having the same energy, focal intensity and pulse duration [53]. This property is one of the reasons why the use of a Bessel Gaussian geometry is interesting in HHG.

From equations (2.1) and (2.5), we can see that the BG beam is characterized by only two parameters: the first is the deviation anglewhich depends on the parameters of the axicon, and the second is the confocal parameter of the Gaussian beam having the same focal intensity and the same pulse duration.

2.1.2.2. Phase Mismatchin the Bessel Gaussian Geometry

Another main advantage of the Bessel Gaussian beam is the possibility to eliminate the phase mismatch in HHG by simply changing the angle of the Bessel cone. According to figure 2.1, the axial wave vector k_z is given by

$$k_z = k_0\cos\gamma \tag{2.6}$$

where k_0 is the plane wave vector.

Substituting $k_z = k_{laser}$ from equation (2.6) into equation $\Delta k_q = qk_{laser} - k_{harmonic}$, one obtains

$$\Delta k_q = qk_z - k_q = q[k_0 - (k_0 - k_z)] - k_q = (qk_0 - k_q) - qk_0(1-\cos\gamma)$$
$$= \left[\frac{2\pi(1-\eta)qP\Delta n}{\lambda} - P\eta N_{at} r_e (q\lambda - q/\lambda) + \text{atomic phase}\right] - qk_0(1-\cos\gamma) \tag{2.7}$$

From equation (2.7), the term $-qk_0(1-\cos\gamma)$ has the same negative sign as the plasma dispersion term. Hence, as long as the ionization fraction η is still less than a critical value, η_{cr}, we can easily choose a suitable value of γ so that the phase mismatch is minimized. This is similar to compensation of the phase mismatch in a waveguide or self guide; however, the range is much higher [27]. In a waveguide, the contribution Δk is proportional to $1/a^2$, where a is the radius of the waveguide.For example, for a= 100 μm,the contribution $\Delta k \approx 36\ m^{-1}$, while for an axicon with a base angle (α) of 2.8^0, the contribution $\Delta k \approx 2337\ m^{-1}$ [27].

2.1.3. Experiment

2.1.3.1. Experimental Setup

A 1 kHz multi-stage, multi-pass, chirped-pulse amplifier system which can produce 2.5 mJ pulses with a duration of 30 fs and centred at 805 nm is used for our experiment. The laser pulses are focused by a combination of a 300 mm focal length lens and an axicon with base angle $\alpha = 0.5^0$, into a 150 mm long gas cell with a glass window at the entrance and a 100 μm pinhole at the exit.The exit angle γ can be estimated from $\gamma \approx (n-1)\alpha = 0.22^0$, where $n = 1.43$ is the refractive index of the axicon. Thus, for the incident wavelength of 805 nm, this axicon can theoretically provide a contribution $k_0(1-\cos\gamma) = \frac{2\pi}{\lambda}(1-\cos\gamma) = 62.5\,m^{-1}$ to the phase mismatch. More details of the experimental setup were described in section 1.2, with the lens now replaced by a lens plus axicon system depicted in figure 2.2. The distance between the lens and the axicon can be varied over the range 10 – 80 mm. The effective peak intensity at the focus is approximately 10^{14}– 5×10^{14} W/cm^2. The pressure inside the gas cell is from 50 to 100 Torr of argon (Ar) gas. The high harmonics generated on-axis or close to the optical axis pass through a 200 nm thick aluminium filter, which removes the fundamental beam. For measurement of the HHG spectrum, a grating incidence spectrometer (GIMS #4 – set point) with a 300 grooves/mm grating is inserted into the beam path. A 13 mm x 13 mm CCD chip (13 μm pixel size, Princeton Instruments PIXIS-XO) at the exit focus plane is used as the detector.

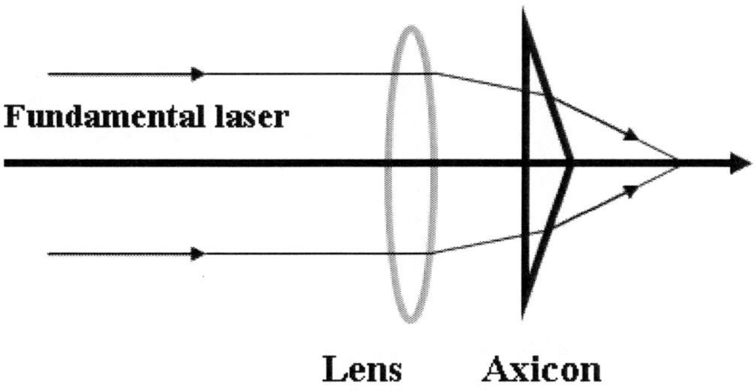

Figure 2.2.Combination of lens and axicon in our experiment.

2.1.3.2. Results and Interpretation

Figure 2.3 shows the HHG spectrum for three different distances between the axicon and the lens. For this measurement, the gas pressure, aperture diameter, pulse energy and focus position of the lens plus axicon system, which is placed close to the exit pinhole and inside the cell, are optimized for maximum total flux of all harmonics. The laser intensity is chosen to be below the saturation intensity. The harmonic generation is phase-matched for a few narrow bandwidth harmonic orders in the cut-off region.

From figure 2.3, it appears that a remarkable increase in the intensity in the cut-off region < 32 nm (harmonic order H > 25) is achieved compared with when only the focusing lens is used (black solid line in figure 2.3). The HHG intensity increases when the distance between the axicon and the lens increases from 1.5 cm to ~ 6.0 cm and then remains constant. A photon flux of ~ 6×10^{12} photons/(cm^2s) for the 27th harmonic (H27) is obtained. The peak intensity of harmonics using the combination of lens and axicon is 3.3 times higher than for the case of using the lens only. This can be explained in terms of compensation of dispersion phase mismatch by changing the cone angle as mentioned in section 2.1.2.2. In our experiment, the variation of the beam size at the axicon by changing the distance between the axicon and the lens is similar to a change of the angle of the axicon cone. This results in a different contribution of the geometric term in the phase mismatch expression (equation 1.11). Therefore, a different harmonic intensity can be obtained as shown in figure 2.3. It is worth noting that during the variation of the distance between the axicon and the lens, the other experimental parameters such as the energy of the laser pulse, the gas pressure, the aperture diameter are kept almost unchanged.

Figure 2.4 shows the dependence of the intensity of the 27th harmonic on the position of the laser focus relative to the exit of the Ar gas cell at pressures of 60 Torr (blue dots) and 95 Torr (green dots).

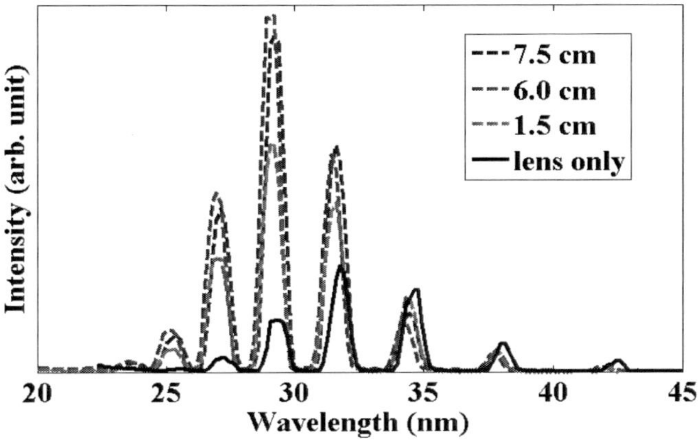

Figure 2.3.HHG spectra for three different distances between the axicon and the lens (7.5, 6.0 and 1.5 cm).The black solid line shows the HHG spectrum for the case without the axicon.

For this measurement, firstly, the aperture diameter is optimized for maximum photon flux and the position of the laser focus is set to z = 0 (close to the exit and inside the gas cell). Then, the laser focus position is varied, while all other parameters are kept constant. Positive values of the interaction length indicate that the focus is deeper inside the gas cell.

Now, let us first recall that in general the qth harmonic field at the exit of the medium with an interaction length L_{med} is the integral of all of the atomic dipoles along the medium and is given by [55]

$$E_q(L_{med}) = \frac{iq\omega}{\varepsilon_0 c} \iint d_q(I,z) \exp\left(\frac{z - L_{med}}{2L_{abs}} \right) \exp\left[i\Delta k_q(I,z)z \right] dz dI \qquad (2.8)$$

where $\Delta k_q(I,z)$ is the phase mismatch between the harmonic radiation and the laser induced polarization and is given by equation (1.11), $d_q(I,z)$ is the amplitude of the high harmonic dipole, L_{abs} is the absorption length, c and ε_0 are the speed of light and vacuum permittivity, ω is the angular frequency of the laser pulse and I is the laser intensity. The integration is over the length of the interaction medium and the intensity envelope of the driving laser field.

Equation (2.8) can be rewritten to express the qth instantaneous harmonic field at the exit of the medium

$$E_q(L_{med}) = \frac{iq\omega}{\varepsilon_0 c}\int d_q(z)\exp\left(\frac{z-L_{med}}{2L_{abs}}\right)\exp\left[i\Delta k_q(z)z\right]dz \tag{2.9}$$

Assuming that when the Rayleigh length is much longer than the interaction length, i.e., over the interaction medium the fundamental laser beam profile is almost unchanged and the generating conditions can be considered to be homogeneous over all the medium, then the phase mismatch term can be expressed as

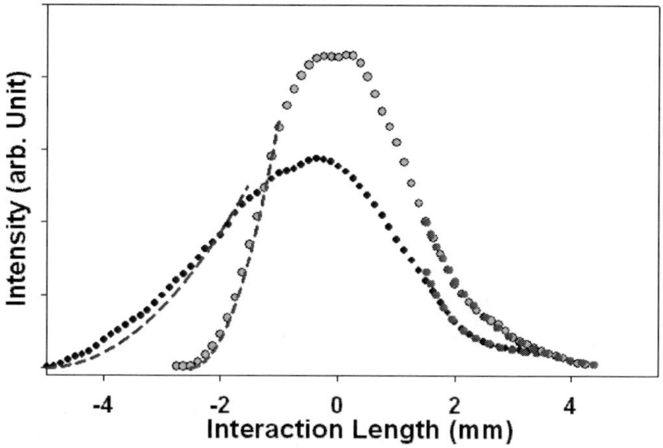

Figure 2.4.Intensity of the 27th harmonic (H27) versus focus position of the lens and axicon system for an argon gas cell at pressures of 60 Torr (blue dots) and 95 Torr (green dots). The quadratic dependence which is theoretically given by equation (2.14) is shown by the red dashed lines. The exponential decay curve which indicates the decrease of the harmonic emission intensity due to re-absorption is shown by the pink dots. Positive values of the interaction length indicate that the focus is inside the gas cell.

$$\Delta k_q = \frac{\pi}{L_{q,coh}} \tag{2.10}$$

where $L_{q,coh}$ is the coherence length of the qth harmonic.

The number of photons emitted on-axis per unit of time and area is proportional to [38, 56]

$$N_{out} \propto \rho^2 \left| E_q (L_{med}) \right|^2 \tag{2.11}$$

where ρ is the gas density.

Substituting equation (2.10) into equation (2.9), and assuming a long Rayleigh length, d_q is independent of z, and therefore equation (2.11) can be expressed as

$$N_{out} \propto \rho^2 d_q^2 \frac{4L_{abs}^2}{1+4\pi^2 (L_{abs}^2 / L_{coh}^2)} \left[1 + \exp\left(-\frac{L_{med}}{L_{abs}} \right) - 2\cos\left(\frac{\pi L_{med}}{L_{coh}} \right) \exp\left(-\frac{L_{med}}{L_{abs}} \right) \right] \tag{2.12}$$

If the absorption of the medium is neglected, i.e., L_{abs} is much larger than L_{med}, then equation (2.12) reduces to

$$N_{out} \propto \rho^2 d_q^2 \frac{4}{4\pi^2 / L_{coh}^2} \left[2 - 2\cos\left(\frac{\pi L_{med}}{L_{coh}} \right) \right]$$

$$\text{or } N_{out} \propto \rho^2 L_{med}^2 d_q^2 \frac{\sin^2\left(\frac{\Delta k_q L_{med}}{2} \right)}{\left(\frac{\Delta k_q L_{med}}{2} \right)^2} \tag{2.13}$$

From equation (2.13), it is clear that when the phase matching condition is achieved and when the influence of the two terms (gas pressure and interaction length) on the harmonic intensity, I_q, needs to be investigated, the harmonic intensity scales as

$$I_q \sim \rho^2 L^2_{med} \tag{2.14}$$

Equation (2.14) is useful to explain figure 2.4.

When a Gaussian beam with diameter of 6.25 mm is focused by a 300 mm lens, the Rayleigh length is ~ 7 mm.From equation (2.5), the Rayleigh length for an associated BG beam produced by an axicon with base angle of 1^0 is ~ 10 mm. Therefore, the Rayleigh length is expected to be longer when a combination of lens and axicon is applied in our experiment.For increasing values of the interaction length, the focus position moves from outside into the gas cell (z < -0.75 mm for 95 Torr and z < -1.5 mm for 60 Torr) and the intensity increases quadratically with interaction length, as shown by the green and blue dots in figure 2.4. The quadratic dependence (red dashed lines fitting curves in figure 2.4) which is theoretically given by equation (2.14) indicates that in our measurement, for these interaction lengths which are smaller than the Rayleigh length, the harmonic emission is phase matched. On the other hand, the harmonic conversion efficiency at 30 nm in our experiment is $\sim 10^{-5}$. The high conversion efficiency and the quadratic scaling of the HHG intensity with the interaction length confirm that the generation process here is phase matched. The phase matching is better when a higher gas pressure is used.The ratio of increasing slopes for the two pressures is ~ 1.5, which is close to the ratio of the gas pressures (95/60 = 1.58).

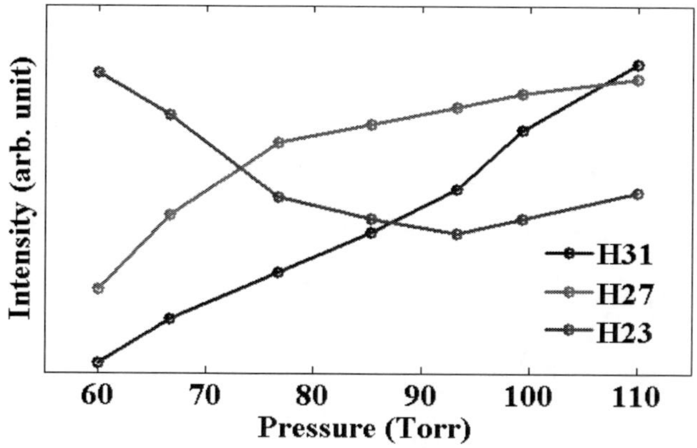

Figure 2.5.Intensity of different harmonic orders versus gas pressure when the distance between the axicon and the lens is 6.0 cm.

It has been shown that the efficiency of the HHG process depends on the absorption of the gas even when the coherence length is infinite [57]. From figure 2.4, we can see that for a longer interaction length the HHG intensity increases slowly and then decreases due to re-absorption in the medium. It is worth reminding here that there is less absorption with the higher order harmonics. The influence of absorption in the medium is shown in figure 2.5 where the intensity of H23 decreases with increasing gas pressure while the intensity of H31 is still increasing as a result of the absorption for H23 in argon being an order of magnitude larger than for H31.In figure 2.4, for $z >$ 1.5 mm the decrease of the harmonic emission intensity with interaction length is dominated by re-absorption and an exponential decay curve (pink dots) can be fitted to the data.

As expressed in equation (2.4), in the case of an ideal Bessel beam, there is no diffraction limit. This leads to a smaller influence of the atomic phase [58]. In our experiment, the reduction of the diffraction limit of the Bessel Gaussian beam created by the combination of the axicon and lens provides a better spatial coherence which is reflected in the very good Gaussian beam profile of the harmonic as shown in figure 2.6. The improvement of the spatial coherence of HHG with the Bessel Gaussian beam compared to that with the Gaussian beam can be estimated from the ratio between the FWHM of the two beam profiles which is 68%.

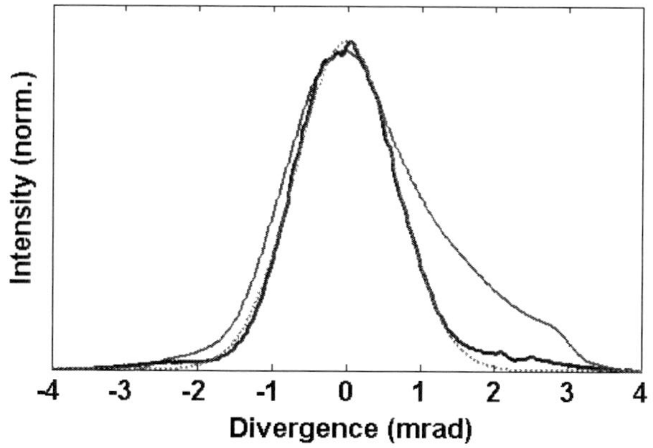

Figure 2.6.Beam profile of the HHG emission for the lens only (red line) and for the lens plus axicon system (blue line) and the theoretical Gaussian profile (green dotted line).

It is apparent that the harmonic beam is more divergent when the fundamental beam is a Gaussian beam only. The better spatial beam profile can create a better collimated harmonic beam and consequently higher brightness.

Conclusion

We have for the first time experimentally investigated the harmonic generation in a semi-finite argon gas cell with a Bessel Gaussian beam by using a combination of a lens and an axicon. Enhancement of the phase matching of a few harmonics in the cut-off region with a higher photon flux (~ 3.3 times higher) and better spatial beam profile (FWHM decreases to 68%) compared to a Gaussian beam. Moreover, in this section, the influence of a combination of lens and axicon, the gas pressure and the absorption effect have been studied.

2.2. Enhancementof High Harmonic Generation Source by Using an Off-Axis Beam

2.2.1. Introduction

Besides enhancement of the conversion efficiency and selection of single harmonics, extension of the cut-off energy to the shorter soft X-ray wavelengths (higher photon energy) is also one of the three main challenges for further High Harmonic Generation experiments [19]. So far, various techniques have been studied to improve harmonic generation in the cut-off region.

Based on the scaling of q_{max} with the ionization potential shown in equation (1.7), a higher harmonic order in the cut-off region can be achieved by using a gas with higher ionization potential, such as helium (24.6 eV) or neon (21.5 eV), as the nonlinear medium. Z. H. Chang et al. [10] claimed that soft X-ray harmonics at an energyof 460 eV (or 2.7 nm) can be observed in helium and 239 eV in neon.

Also, according to equation (1.7), E. Gibson et al. [59] demonstrated the generation of very high order harmonics up to 250 eV by using ionized argon.

The ionization potential (I_p) of ionized argon (27.6 eV) is much higher than that of neutral argon (15.8 eV).

Using a longer wavelength driving laser is also one of the techniques which are able to significantly extend the cut-off photon energy in high harmonic generation. From equation (1.5), the pondermotive energy (U_p) is linearly proportional to λ^2. Thus, a higher kinetic energy (U_p), i.e., a higher cut-off photon energy can be obtained by applying longer wavelength light as the fundamental beam. T. Popmintchev et al. [60] showed that the phase matched harmonic generation can be extended up to extremely high photon energies (around 330 eV) by using mid-infrared pulses with a wavelength of 1.3 μm as the fundamental beam.

Another interesting technique was studied by J. Carrera et al. [61]. In their theoretical work, by using intense few-cycle chirped laser pulses and through the optimization of the chirping rate parameters, a significant extension of the cut-off photon energy can be achieved up to around 375 eV (or 3.3 nm). This can be explained in terms of the modification of the trajectory of the electron wave packet by chirping the laser and when a short trajectory is dominant the cut-off wavelength can be shortened.

However, each of these techniques mentioned above has some limitation. For example, an atom with higher ionization potential like helium has an exceptionally small effective nonlinearity, which leads to a very low photon flux. When using a longer wavelength, it is theoretically and experimentally proven that the single-atom yield scales as $\lambda^{-5.5\pm0.5}$ [62]. Hence, this strongly reduces the HHG efficiency. In an ionized medium, the HHG efficiency is limited by plasma induced defocusing and increased phase mismatch at high ionization levels. For optimizing the chirp parameters presented by J. Carrera et al. [61], only a purely single-atom response has been investigated theoretically.

In this section, we propose a new experimental approach in which an off-axis laser beam is used to control the phase matching condition in the HHG with a strong femtosecond laser field in a semi-infinite gas cell. Different from the use of a combination of axicon and lens to vary the geometry phase mismatch term for compensation of the total phase mismatch which was demonstrated in section 2.1, an off-axis beam can lead to an enhancement or destruction of the phase matching which is due mostly to the variation of the dipole phase mismatch term in the presence of the second off-axis beam. Through controlling the phase matching condition the harmonic emission can be suppressed or improved. A cut-off photon energy of up to 220 eV (~ 5.5 nm in wavelength) is successfully observed in our experiments when the off-axis beam is applied.

2.2.2. Influence of the Off-axis Beam on the Fundamental Field Generating HHG

The key to the QPM technique using a relatively weak counter-propagating beam as well as the perpendicular beam is that the counter-propagating field creates a phase modulation of the fundamental field in the overlap region [47, 48, 49]. Thus, in this section, we investigate the influence of the off-axis beam used in our experiment on the fundamental field.

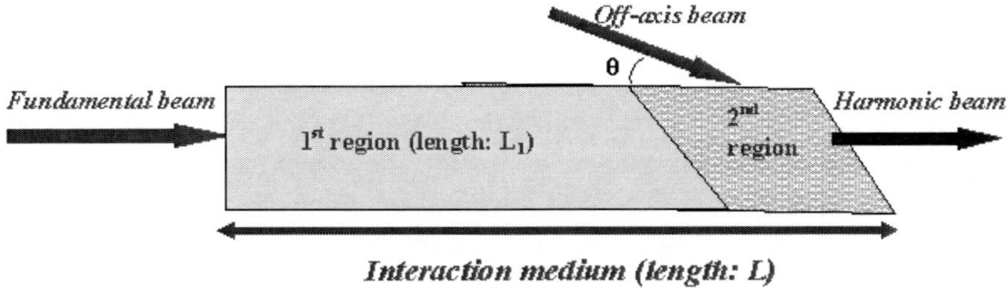

Figure 2.7.Scheme of the off-axis beam setup in our experiment.

The fundamental beam is focused onto atoms with a total interaction length of L. Then, a relatively weak off-axis beam intersects the fundamental beam at an angle θ (θ is very small in this setup) leading to a second region (overlapping region) which is different to the first region. In contrast to the setup with the counter-propagating beam shown in figure 1.6, the length of the second region is only a fraction of the total interaction length. The overall interaction length and the length of the second region can be varied.

From figure 2.7, it appears that the presence of the off-axis beam does not influence the HHG field in the first region and only changes the HHG field in the second region. However a small variation of the HHG field in the second region can result in a significant change in the overall HHG signal within the whole interaction medium.

Now, we will investigate the influence of the off-axis field on the fundamental field in the second region.

We can write

$E_1 e^{i(kz-\omega t)}$:The fundamental field to generate HHG.

$E_2 e^{i(kz\cos\theta + ky\sin\theta - \omega t)}$: The influence of the off-axis field in the plane of the fundamental field (plane of HHG) [49], where θ is the angle between the two fields.

The sum of the two fields can be written as

$$E = E_1 e^{i(kz-\omega t)} + E_2 e^{i(kz\cos\theta + ky\sin\theta - \omega t)} = E_s e^{i\varphi} \tag{2.15}$$

From equation (2.15), one can obtain

$$\varphi = \tan^{-1}\{\frac{E_2 \sin(kz\cos\theta + ky\sin\theta - kz)}{E_2 \cos(kz\cos\theta + ky\sin\theta - kz) + E_1}\} + (kz - wt) \tag{2.16}$$

$$\text{Setting } \phi = \tan^{-1}\{\frac{E_2 \sin(kz\cos\theta + ky\sin\theta - kz)}{E_2 \cos(kz\cos\theta + ky\sin\theta - kz) + E_1}\} \tag{2.17}$$

from equations (2.16) and (2.17), equation (2.15) becomes

$$E = E_s e^{i\phi} e^{i(kz-wt)} \tag{2.18}$$

In general, comparing the form of the sum field E in equation (2.18) and the form of the fundamental field generating HHG, we find that the phase factor is $e^{i\phi}$. In the case of counter-propagating beams ($\theta = 180^0$), the phase factor $\phi = -\tan^{-1}\dfrac{E_2 \sin 2kz}{E_1 + E_2 \cos 2kz}$. This additional phase factor of the fundamental field results in significant disruption to the total phase matching condition, i.e., to turn off the harmonic signal as observed by S. L. Voronov et al. [48].

However, in our configuration, the angle θ is very small (~ 5^0), thus $\cos\theta \sim 1$ and $\sin\theta \sim 0$. If we define $kz\cos\theta + ky\sin\theta - kz = \gamma$, then $\gamma \to 0$.

Consequently

$$\phi = \tan^{-1}\{\lim_{\gamma\to 0}\{\frac{E_2 \sin\gamma}{E_2\cos\gamma + E_1}\}\} = \tan^{-1}\{\lim_{\gamma\to 0}[\frac{(E_2/E_1)\sin\gamma}{1+(E_2/E_1)\cos\gamma}]\}$$

$$= \tan^{-1}\{\frac{\{(E_2/E_1)\lim_{\gamma\to 0}\frac{\sin\gamma}{\gamma}\}\gamma}{\lim_{\gamma\to 0}[1+(E_2/E_1)\cos\gamma]}\} = \tan^{-1}\{\frac{(E_2/E_1).0}{1+(E_2/E_1)}\} = 0 \tag{2.19}$$

This means there is no phase factor which indicates that in our experiment the off-axis beam does not change the phase of the fundamental field like in previous studies [47, 48, 49].

Consider the amplitude of the sum of the field, E_s, in the second region given by

$$E_s = \sqrt{E_1^2 + E_2^2 + 2E_1E_2} = E_1 + E_2 \tag{2.20}$$

Equation (2.20) indicates that the off-axis beam only changes the amplitude of the fundamental field in the overlapping region by adding its amplitude. This allows us to change the phase matching condition leading to control of the high harmonic generation which we will present in more detail in the next section.

2.2.3. Influence of theOff-axis Beam on the HHG field

We refer to equation (2.8) which gives the qth harmonic field at the exit of the medium with interaction length L_{med}. In this equation, the amplitude of the high harmonic dipole, $d_q(z)$, is dependent on the intensity of the driving laser field with peak intensity I_0 which varies along the focusing direction (z) and perpendicular to the focusing direction across the beam profile, and is also dependent on properties of the gas medium such as the ionization energy and orientation of the atoms or molecules relative to the driving field. In the single-atom response model, d_q is estimated as [38]

$$d_q \approx I^3 \tag{2.21}$$

in the intensity window of the driving pulse I to I+dI.

The presence of the off-axis beam divides the interaction medium into two different regions shown in figure 2.7.

In the case without the off-axis beam, equation (2.8) can be rewritten as the sum of two fields generated in two homogeneous regions

$$E_q^{(1)}(L_{med}) = E_{q1}^{(1)}(L_1) + E_{q2}^{(1)}(L_{med} - L_1)$$
$$= \frac{iq\omega}{\varepsilon_0 c}\exp\left(-\frac{L_{med}}{2L_{abs}}\right)\left[\int\int_0^{L_1} d_q(I,z)\exp\left(\frac{z}{2L_{abs}}\right)\exp[i\Delta k_q(I,z)z]dzdI \right. \quad (2.22)$$
$$\left. + \int\int_{L_1}^{L_{med}} d_q(I,z)\exp\left(\frac{z}{2L_{abs}}\right)\exp[i\Delta k_q(I,z)z]dzdI\right]$$

When the off-axis beam interacts with the fundamental beam and overlaps in time, the laser intensity only substantially increases within the overlapping region (the second region) as indicated in equation (2.20). As a result, in this region, there is not only an abrupt increase of the instantaneous response part of the refractive index which follows the intensity profile of the laser field ($\Delta n_{inst} = n_2 I$) but also a higher ionization fraction (η). This leads to the situation where both the first term (medium dispersion) and the second term (plasma dispersion) of the total phase mismatch, which is given by equation (1.11), suddenly change in the second region. In addition, the fourth term, i.e., the atomic dipole phase, strongly depends on the laser field which generates the HHG [63]. Due to the sudden variation of the total phase mismatch which originates from the change of the first, second and fourth term contribution in the overlapping region, the overall q^{th} harmonic field at the exit of the medium in equation (2.22) can now be described as

$$E_q^{(2)}(L_{med}) = E_{q1}^{(2)}(L_1) + E_{q2}^{(2)}(L_{med} - L_1) =$$
$$\frac{iq\omega}{\varepsilon_0 c}\exp\left(-\frac{L_{med}}{2L_{abs}}\right)\left[\int\int_0^{L_1} d_q(I,z)\exp\left(\frac{z}{2L_{abs}}\right)\exp[i\Delta k_q(I,z)z]dzdI \right.$$
$$\left. + \int\int_{L_1}^{L_{med}} d_q'(I,z)\exp\left(\frac{z}{2L_{abs}}\right)\exp[i\Delta k_q'(I,z)z]dzdI\right]$$

(2.23)

Since there is no change in the first region, $E_{q1}^{(2)}(L_1)$ would be equal to $E_{q1}^{(1)}(L_1)$. In the overlapping region, when there is the off-axis beam, the total phase mismatch becomes $\Delta k_q^{'}$ and the amplitude of the harmonic dipole changes to $d_q^{'}(z)$.

In equations (2.22) and (2.23), the superscript (1) indicates that only the fundamental beam generates HHG and the superscript (2) indicates the presence of the off-axis beam, while the subscript (1) indicates the harmonic field in the first region and the subscript (2) indicates the harmonic field in the second region.

Destruction or enhancement of the harmonic emission can be achieved through equation (2.23). If the total phase mismatch is changed, the harmonic fields in the first region and the second region can be subtracted or added depending on their relative phase. Hence, the overall generated field given by $E_q^{(2)}(L_{med}) = E_{q1}^{(2)}(L_1) + E_{q2}^{(2)}(L_{med} - L_1)$ can be varied, i.e., the harmonic signal can be turned off or enhanced. Since the phase of the harmonic field shown in equation (2.8) only depends on the total phase mismatch Δk_q, we do not need to consider the variation of the amplitude of the harmonic dipole $d_q(z)$ in this investigation.

In the next section, we will describe experiments using the off-axis beam to turn off or enhance the HHG in the semi-infinite gas cell. Also, based on these experimental results, we are able to determine which term of the total phase mismatch plays a dominant role in controlling the harmonic emission in this configuration.

2.2.4. Experiment

2.2.4.1. Experimental Setup

The laser system and the harmonic production and detection systems used in this experiment are the same as those presented in section 1.2. The laser beam is split into two beams with an intensity ratio of first beam to second beam ~ 1:4. The two laser beams are focused to the same spot in the gas-filled cell at a small angle of ~ 5^0. The first laser beam with low intensity is used as the off-axis beam and the second is used for generation of harmonics. To produce a time delay between the two beams, a stepping motor with a spatial resolution of 25 nm is installed. Positive time delay means that the first beam precedes the second beam. The gases argon and helium have been used as the nonlinear media. Figure 2.8 shows the arrangement of the two beams.

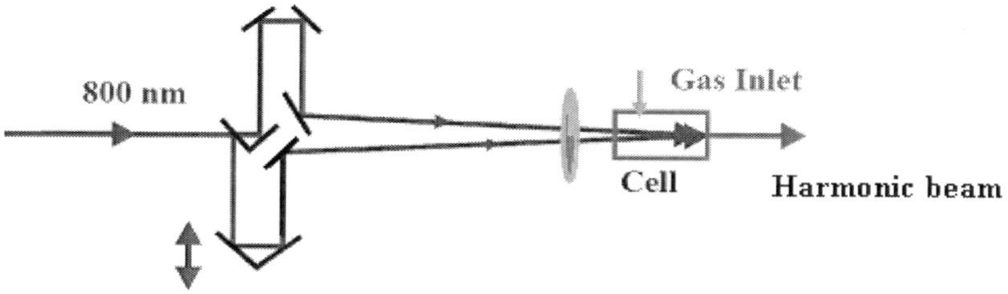

Figure 2.8. Two beam scheme.

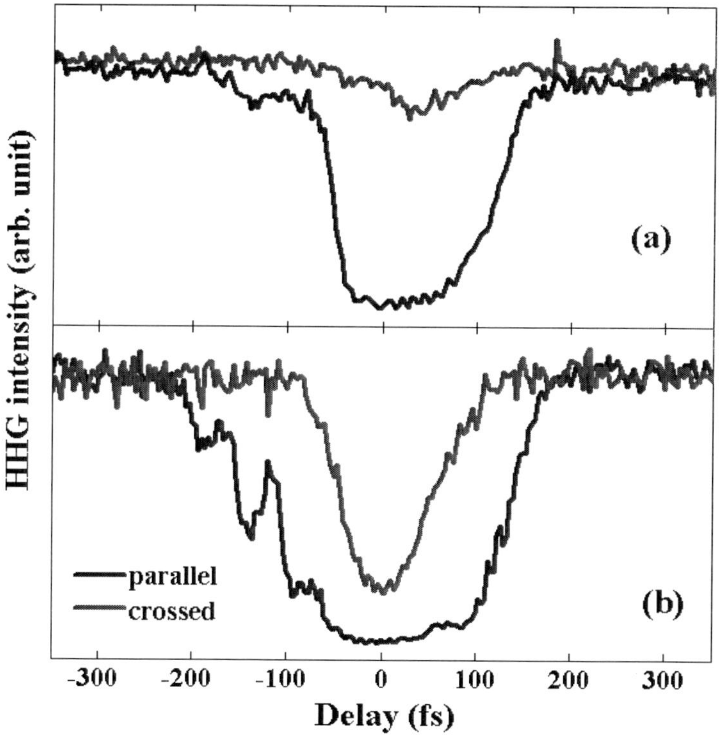

Figure 2.9. Variation of the intensity of harmonic H23 from argon gas versus delay between the two laser pulses. The polarizations of the two beams are parallel (blue) and crossed (red). The intensity of the off-axis beam is (a) 1×10^{13} W/cm^2 and (b) 3×10^{13} W/cm^2.

2.2.4.2. Results and Interpretation

Firstly, we show the use of an off-axis beam to turn off the harmonic emission. In this experiment, the delay of the off-axis beam is set to a very long delay (e.g. ~ 1 ps) so that there is no influence of the off-axis beam on the second beam. At this point, the pressure, aperture diameter, focus position of the lens, which is placed close to the exit and inside the gas cell, and chirping of the fundamental beam are optimized for harmonic emission close to the cut-off region. This means that the harmonics are in the phase matching condition along the axis of the second beam. For argon gas, the total interaction length (L_{med}) is approximately 4 mm. Then, the harmonic intensity is scanned following the delay between the two beams. When the two beams overlap in time, the spatial overlapping region is less than 1 mm. Figure 2.9a depicts the intensity of the harmonic H23 from argon versus the time delay when the intensity of the first pulse is ~ 10^{13} W/cm^2. A similar dependence is observed for other harmonics ranging from H17 to H27 and also for the harmonic spectrum from helium.

In figure 2.9a, it is apparent that when the polarizations of the two laser fields are parallel, the harmonic signal is dramatically suppressed around zero delay time. This phenomenon is expected as the overall electric field of harmonic H23 is eliminated due to the presence of the off-axis beam. In order to make it more clear, it is shown that when two beams overlap in time:

Due to no change in the first region, the harmonic field in this region remains the same $E^{(2)}_{23(1)}(L_1) = E^{(1)}_{23(1)}(L_1)$.

In the second region, as theoretically explained in section 2.2.3, owing to the presence of the off-axis beam which leads to the variation of the total phase mismatch, the harmonic field in this region becomes $E^{(2)}_{23(2)}(L_{med} - L_1)$. By varying the overlapping length and the intensity of the off-axis beam, the relative phase of $E^{(2)}_{23(2)}(L_{med} - L_1)$ compared to that of the harmonic field in the first region can be controlled through change of the total phase mismatch. At a particular point, these two harmonic fields are strongly subtracted from each other and hence the harmonic signal is turned off.

When the polarizations of the two laser fields are crossed the change of harmonic intensity is small (figure 2.9a). Comparing the changes of the harmonic intensity for two cases of the polarizations and since the ionization rate of the atomic medium in the overlapping region under the influence of the first pulse should be independent of the polarization, the contribution of the plasma dispersion to the change of the total phase mismatch (second term of equation 1.11) does not need to be considered. In addition, for the weak off-axis beam, the instantaneous response of the medium involving the nonlinear refractive index which follows the intensity profile of the fundamental laser field can be ignored. This means that the variation of the medium dispersion phase mismatch in the presence of the off-axis beam is also negligible. Therefore, the change of the total phase mismatch will be mainly caused by the dipole phase. The interaction of the second field with the free electrons influences the trajectory and re-scattering of the electrons causes a large change in the dipole phase. Additional evidence for the dominant contribution of the harmonic dipole phase to suppression of the harmonic emission is the different response of the harmonic intensity on the positive and negative sides close to zero time delay (figure 2.9a). For negative delay, the generating beam precedes the off-axis beam, so that then when the interaction between two beams occurs, the electrons have already been in the continuum state. Conversely, for positive delay, this interaction occurs when the electrons are still in the ground state leading to a slower response.

When a more intense first laser field (~ 3×10^{13} W/cm^2) is applied, as shown in figure 2.9b, the time profile of the turn-off region is much broader than the pulse duration for the case of parallel polarizations because the response of the free electron is very sensitive to the external field and therefore the small field in the tail of the first pulse can disturb the trajectory of the free electrons. Furthermore, for the high intensity first field, the contribution of the instantaneous response of the nonlinear refractive index needs to be considered and the variation of the HHG intensity for crossed polarizations is similar to the intensity profile of the fundamental laser field.

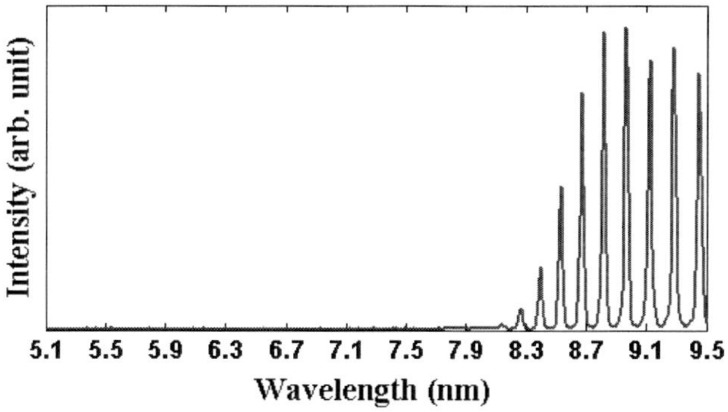

Figure 2.10. Typical optimized HHG spectrum from helium gas in the cut-off region generated in the semi-infinite gas cell with no off-axis beam.

To increase the generated photon energy an increase of the laser intensity is required, but then the ionization rate will be higher causing significant plasma dispersion. The total phase mismatch is large and therefore the output harmonic field is very weak. In this case the off-axis beam can be used to enhance the higher order harmonic emission where the contribution of short quantum paths is high.

Firstly, we display a typical optimized harmonic spectrum from helium gas in the cut-off region generated in our semi-infinite gas cell using only the fundamental laser beam.

From figure 2.10, it appears that in our traditional configuration the shortest harmonic wavelength that can be obtained is ~ 8.3 nm.

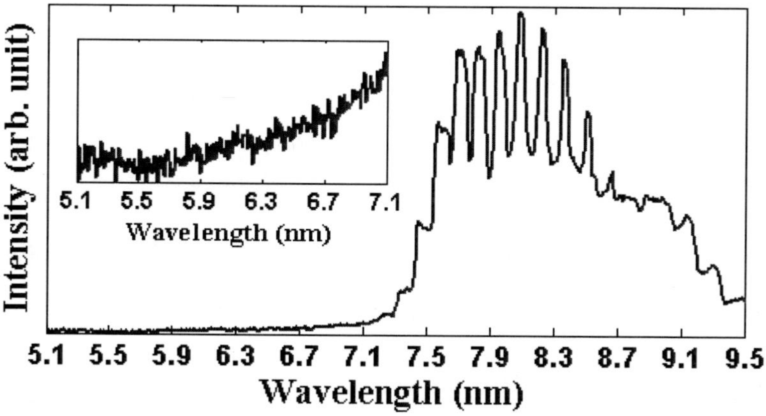

Figure 2.11.HHG spectrum from heliumgas with the off-axis beam at approximately zero delay.The inset shows an expansion in the wavelength range 5.1 nm to 7.1 nm.

When the off-axis beam is used to create the new configuration shown in figure 2.7, there is an enhancement of harmonics with very high photon energy from helium gas at 400 Torr, as shown in figure 2.11. We note that the optimized experimental conditions such as the gas pressure, the chirp and the intensity of the fundamental laser beam are different from the optimal conditions for the typical harmonic spectrum shown in figure 2.10. The same enhancement effect is seen for HHG from argon gas but the generation of very high harmonic orders from helium is more interesting. The variation of the harmonic spectrum with change of delay time between the two fields is shown in figure 2.12. The influence of the off-axis beam is strong for delay times of 0 and ± 30 fs when the intensity of the off axis beam is high and a large variation of the dipole phase is expected. With an off-axis beam, harmonics can be generated down to 7.3 nm and a second cut-off with low flux occurs down to ~ 5.5 nm (figure 2.11 and inset). Since we are not able to achieve phase matching for harmonics with wavelength less than 7.3 nm leading to very low photon flux of these harmonics, clear peak intensities attributed to harmonic emission cannot be observed in this range.

In our experiment, if the off-axis beam is blocked, there would be almost no HHG signal in the wavelength range less than 8 nm. This means that for harmonics in this range, the overall electric field given by equation (2.8) must be nearly eliminated due to the large total phase mismatch in the whole interaction medium. We can rewrite

$$E_q^{(1)}(L_{med}) = E_{q1}^{(1)}(L_1) + E_{q2}^{(1)}(L_{med} - L_1) \to 0.$$

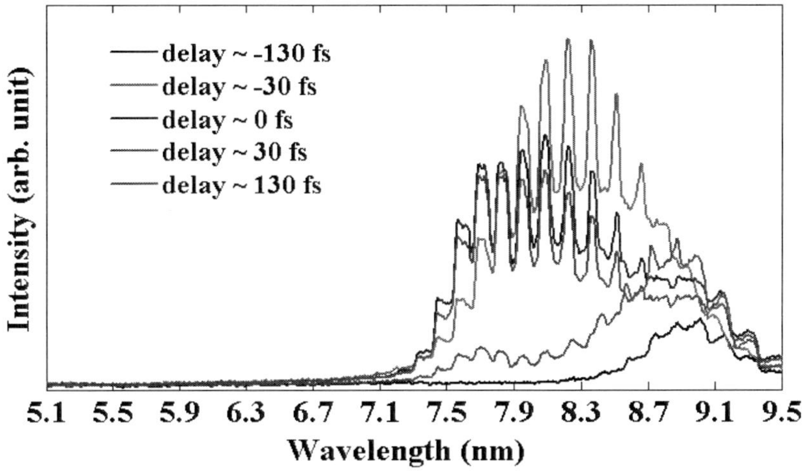

Figure 2.12. HHG spectrafromhelium gas with different time delays between the two beams.

When there is an off-axis beam, at zero time delay, as explained above, the total phase mismatch in the overlapping region (second region) varies due mostly to the change of the dipole phase. Thus, the harmonic field in the second region is modified from $E_{q2}^{(1)}(L_{med} - L_1)$ to $E_{q2}^{(2)}(L_{med} - L_1)$ and is added to the harmonic field in the first region ($E_{q1}^{(2)}(L_1)$). Consequently, the overall harmonic field at the exit of the medium $E_q^{(2)}(L_{med}) = E_{q1}^{(2)}(L_1) + E_{q2}^{(2)}(L_{med} - L_1)$ can be strongly enhanced, i.e., the harmonic emission in the cut-off region is extended and the HHG signal can be found down to 5.5 nm.

Conclusion

We have demonstrated the use of an off-axis laser beam to alter the phase mismatch leading to suppression or enhancement of the HHG process. With an off-axis beam harmonics around 7.3 nm have been generated and a harmonic signal with low flux down to 5.5 nm has been observed. In addition, in this study, the importance of the harmonic dipole phase on the phase mismatch has been investigated.

Chapter 3

INFLUENCE OF DRIVING LASER ON SPECTRAL FEATURES OF HIGH HARMONIC GENERATION

3.1. INTRODUCTION

In general, High Harmonic Generation in noble gases can be produced by focusing laser pulses with an intensity of 10^{14}-10^{15} W/cm^2 into the nonlinear medium. As mentioned in section 1.1.3, to investigate the HHG process, one needs not only to consider the single-atom response but also macroscopic effects. Therefore, when the laser intensity is varied, the variation of the single-atom dipole moment obtained by solving the Schrödinger equation and the variation of the propagation equation of the harmonic field in the nonlinear medium influence both the strength and spectrum of the harmonic generation [7, 64]. It has been shown that the harmonic dipole $d_q(t)$ has a laser intensity dependence that may be obtained from a quantum mechanical theory [21, 63, 64]. To simplify the calculation of $d_q(t)$, a similar result has been proposed with a power law $d_q \sim I^3$by averaging over the quantum interference, where I is the driving laser intensity [38]. On the other hand, in relation to the macroscopic response, the propagation equation of the harmonic field created in the medium depends strongly on the phase mismatch between the fundamental laser field and the harmonic field which is mainly affected by ionization effects when the laser intensity is varied [7, 64]. In these papers, the authors also studied the laser intensity dependence of the harmonic strength in detail both in theory and experiment. The laser intensity dependence of the number of photons can be described by the power law I^q, where q is the harmonic order and I is the laser intensity for the intensity region below the saturation intensity. The behaviour of the harmonic strength above saturation is more complex and cannot be described by the simple power law I^q. In this region, because of the rapid increase of the ionization fraction, macroscopic effects such as the depletion of the neutral atom population and the increase of the free electron density need to be considered.

The influence of the ionizing gaseous medium on the blue-shifting and broadening of the fundamental spectrum has been studied in [65, 66]. During the ionization, a rapid increase of the free electron density induces a variation of the gas refractive index, which results in a shift of the wavelength of the fundamental beam. For an incident wavelength λ_0 and an interaction length L, this spectral shift $\Delta\lambda$ is given by [66]

$$\Delta\lambda = \frac{e^2 N_i \lambda_0^3 L}{8\pi^2 \varepsilon_0 m_e c^3} \frac{d\eta}{dt} \tag{3.1}$$

where N_i is the ion density and η is the ionization rate. An estimate of the shift depends on an accurate calculation of the ionization rate.

The spectral shift of the fundamental beam in turn leads to a spectral shift of the particular harmonic q given by [67, 68, 69]

$$\Delta\lambda_q = \Delta\lambda / q \tag{3.2}$$

The harmonic spectrum can also be shifted due to the refractive index change at λ_q caused by the presence of the free electrons. However, this shift is much smaller than the previous one [68, 69].

Another interesting theoretical study of the spectral structure of the high harmonic emission from atomic gases was conducted by Kan et al. [70]. By looking at the single-atom response, they showed that the splitting and blueshift of the harmonic spectrum is due to the intensity or time dependent phase shift of the dipole acceleration that is responsible for harmonic generation. There are two intensity-dependent phase terms, which are the classical accumulated phase of the electron wave function of the qth harmonic and q times the returning phase. When the intensity of the fundamental beam changes during the laser pulse, the relative contribution from these two terms to the total harmonic phase will change depending on the electron trajectory in the continuum states. The different increase of the harmonic phase with the fundamental laser intensity leads to a blueshift and splitting of the harmonic spectrum. It is useful to recall that under these model pulse conditions, where the ground state is depleted just before the peak of the pulse, the harmonics are generated only on the leading edge of the fundamental pulse.

Another theoretical approach to study the intensity dependence of the harmonic emission based on the single-atom response is to consider the harmonic signal produced on the two different sides of the laser pulse [71]. It has been shown that a temporal modulation of the harmonic signal due to a temporal modulation of the instantaneous harmonic frequency is mapped onto the frequency domain and this leads to a spectrally broadened harmonic pulse with a blueshifted leading edge and a redshifted trailing edge [71].

Recently, the contribution of interference between the quantum paths to the harmonic emission has been theoretically and experimentally investigated [72, 73]. As we know, there are typically two electron trajectories leading to the emission of radiation at the same frequency but with different temporal and spectral properties: a short trajectory corresponding to a short return time and to a slow associated dipole phase variation and a long trajectory corresponding to a return time close to one optical period and to a rapid associated dipole phase variation [19]. At low intensity, basically there is only a single quantum path contributing. Conversely, at high intensity, both quantum paths participate in the harmonic generation. The fast regular oscillations of the intensity dependent harmonic signal and the interference fringes in the harmonic spectrum due to the quantum path interferences have been theoretically observed by A. Zair et al. [72, 73]. In addition, it has been shown that the contribution of different electron trajectories to the harmonic radiation can induce a splitting and frequency shift of the harmonic lines with the central harmonic frequencies produced mainly by the shortest trajectories and the shifted frequencies by the longer trajectories [74].

By studying the propagation equation of the harmonic field, the effects of ionization on the high harmonic spectral broadening and splitting have also been reported [75, 76, 77]. The increase of the ionization rate due to an increase of the driving laser intensity causes a rapid decrease of the neutral gas density and an increase of the free electron density. Both of these affect the propagation equation and modulate the temporal profile of the harmonic signal. The modulation of the envelope of the harmonic field in the time domain causes a splitting of the harmonic line in the frequency domain.

We discuss the influence of the fundamental beam on the spectral features of the harmonics. The macroscopic phase matching which plays a particularly significant role in the harmonic radiation using a long gas cell and the interplay between the single-atom response and the macroscopic response are analyzed. In addition, based on these considering, we can determine the conditions for the production of sharp and strong harmonics.

3.2. Theoretical Background

A physical understanding of the HHG process is provided by the three-step model where the interaction between the laser field and the atoms is separated into ionization, acceleration and recombination [3, 8]. A full quantum mechanical theory which recovers the semi-classical model and considers quantum effects such as tunneling, diffusion and interference has been given within the strong field approximation (SFA) model to describe aspects of the HHG process more precisely [21]. In the SFA theory, the HHG from a single atom can be obtained by calculating the dipole acceleration of a returning electron which has gained momentum in the presence of the oscillating electric field. However, in general, an understanding of the HHG process based on only the single-atom dynamics is not sufficient to explain the experimental data completely. In order to achieve good agreement between theory and experiment, one must also take into account the effects of propagation and phase mismatch between the harmonic field and the fundamental field in the macroscopic medium [7, 64]. In this section, we discuss the influence of the fundamental laser intensity on the spectral features with consideration of the macroscopic effects.

3.2.1. Influence of Driving Laser Intensity on the Harmonic Yield

First, let us recall that if we assume a constant nonlinear polarization strength and homogeneous generating conditions within a short interaction length L_{med} and neglect absorption, equation (2.13) allows us to estimate the number of emitted photons on-axis per unit time per unit area for the qth harmonic and equation (2.21) provides an estimate of the amplitude of the harmonic dipole $d_q \approx I^3$. The intensity of the qth harmonic per unit area is then the total harmonic signal, I_q, at the end of the pulse and is expressed as

$$I_q \propto L_{med}^2 \int_0^{\infty} I^6 \frac{\sin^2\left(\frac{\Delta k_q L_{med}}{2}\right)}{\left(\frac{\Delta k_q L_{med}}{2}\right)^2} dI \tag{3.3}$$

where I is the envelope intensity of the laser pulse which for a Gaussian intensity profile can be defined in terms of the peak laser intensity I_0

$$I = I_0 \exp(-t^2/\sigma^2) \tag{3.4}$$

where $\sigma = \tau/(2\sqrt{\ln 2})$ is the temporal width with respect to the FWHM τ.

Moreover, in equation (3.3), Δk_q is the total phase mismatch between the fundamental field and the harmonic field and can be calculated from the sum of four terms: the neutral gas dispersion phase mismatch, the plasma dispersion phase mismatch, the geometric phase mismatch and the atomic dipole phase mismatch, as indicated in detail by equation (1.11). When overall phase matching is achieved, the dependence of the harmonic intensity on the position of the laser focus relative to the exit of the gas cell has been studied in chapter 2 and it appears that for an interaction length smaller than the Rayleigh length the harmonic intensity increases quadratically with the interaction length. With the use of a 300 mm focal length lens, the Rayleigh length is around 7 mm, and thus the interaction length in this case should be less than 5 mm.

In general, the variation of the instantaneous intensity in time and in space, I, causes a change of the atomic dipole phase. As stated in section 1.1.3.2, the origin of this phase, φ_q, is the action acquired by the electron wave packet along the trajectory leading to the emission of the qth harmonic in the continuum state. To a first approximation this phase scales linearly with the laser intensity, $\varphi_q = -\alpha_q I$, where α_q is a coefficient relating the electron trajectories ($\alpha_q \approx 1-5\times10^{-14}\ cm^2/W$ for the short trajectory and $20-25\times10^{-14}\ cm^2/W$ for a long trajectory [73, 78]). Because this phase can change with the laser intensity it may strongly influence the phase matching to the extent that the laser intensity varies within the generating medium. The contribution to the total phase mismatch on-axis of the

q^{th}harmonic induced by the variation of the laser intensity along the direction of propagation, $\nabla\varphi_q = -\alpha_q \frac{\partial I}{\partial z}$ is given by equation (1.30). During the laser pulse, I and therefore $\nabla\varphi_q$ vary with time. Moreover, the variation of the laser intensity with time modifies the neutral gas density and the plasma density leading to a change of the neutral gas dispersion and plasma dispersion. Consequently, according to equation (3.3) a variation of the fundamental laser intensity influences the harmonic intensity through a change of Δk_q and the amplitude of the harmonic dipole, d_q.

3.2.2.Influence of Modulation of the Harmonic Dipole Phase on the Spectrum of HHG

It was indicated that the change of the dipole phase resulting from the variation of the fundamental laser intensity affects the total phase mismatch leading to a change of the harmonic intensity. We discuss there theoretically the effects of the modulation of this intensity dependent phase on the spectrum of the HHG.

The variation in time of the fundamental laser intensity, I(t), induces a change in the instantaneous frequency of the harmonic emission through a harmonic phase modulation process [71, 73]. The instantaneous harmonic frequency during the pulse is given by [71, 73]

$$\omega_q(t) = q\omega + \alpha_q \frac{\partial I}{\partial t} \tag{3.5}$$

Here $q\omega$ is the central harmonic frequency and the term $\alpha_q \frac{\partial I}{\partial t}$ gives the variation of the instantaneous frequency with time. If the signal at precisely the harmonic central frequency always corresponds in time to the central part of the laser temporal envelope, then the emission from the envelope edge is shifted to different side frequencies: a redshift is produced on the trailing edge and a blueshift on the leading edge. This phenomenon can be understood as follows. For the leading edge of the pulse, $\frac{\partial I}{\partial t} > 0$, i.e., the term $\alpha_q \frac{\partial I}{\partial t}$ is positive and therefore from equation (3.5) a higher frequency harmonic signal is emitted. Conversely, for the trailing edge of the pulse, $\frac{\partial I}{\partial t} < 0$, i.e., the term $\alpha_q \frac{\partial I}{\partial t}$ is negative and a lower frequency harmonic signal is emitted. Physically, we may explain this feature in the following way. Electrons that are ionized during the few cycles immediately before the peak of the pulse experience an additional acceleration before returning and this leads to a blueshifted harmonic spectrum. Conversely, a redshift would be visible when electrons are ionized after the peak of the pulse and experience a deceleration [79].

The mapping process is illustrated in figure 3.1. When the harmonic emission is confined to the central part of the light pulse, the mapping process is almost linear. From equation (3.5), the spectral broadening is given by

$$\Delta\omega_q(t) = \alpha_q \frac{\partial I}{\partial t} \tag{3.6}$$

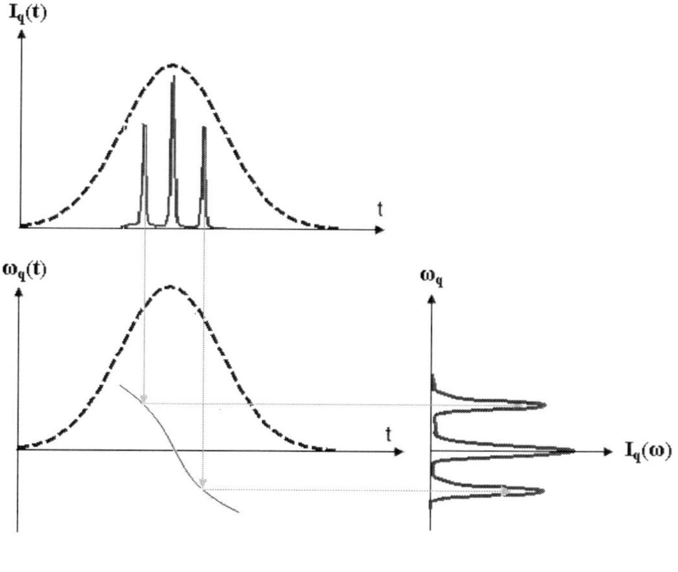

Figure 3.1. The mapping process of the qth harmonic order: The temporal modulation of the harmonic signal I_q (t) is mapped onto the frequency domain as a result of the temporal modulation of the instantaneous harmonic frequency $\omega_q(t)$. The dashed lines represent the intensity profiles of the fundamental.

For a Gaussian intensity profile, during the laser pulse, the phase matching is transiently optimized at certain intensities I_n at given times t_n depending on the peak intensity, I_0 [71]. Similarly to equation (3.4), we can write

$$I_n = I_0 \exp(-t_n^{\,2}/\sigma^2) \tag{3.7}$$

From equations (3.6) and (3.7), the instantaneous harmonic frequency at the laser intensity I_n given by equation (3.5) can be calculated as a function of the scanning intensity, i.e., the peak intensity, I_0 [71]

$$\omega_q(I_0) = q\omega \pm \alpha_q \frac{I_n}{\sigma} 2\sqrt{\ln\left(\frac{I_0}{I_n}\right)} \tag{3.8}$$

Since a long path with very large coefficient, α_q, is predominantly phase-matched off-axis, from equation (3.8) it is found that the off-axis region would give a larger shifting and broadening than the on-axis region.

3.2.3. Influence of the Harmonic Propagation on the Spectrum of HHG

A variation of the fundamental laser intensity during the laser pulse causes not only a change of the harmonic dipole phase that was studied in section 3.2.2 but also a change of the ionization rate of the medium. Basically, the total ionization fraction produced on the trailing edge of the pulse is higher than that produced on the leading edge which results in different phase mismatches between the fundamental field and the harmonic fields emitted a few cycles before and after the peak of the pulse. This influences the propagation of the harmonic field and may lead to modification of the spectral features.

For the qth harmonic, the Maxwell equation of the harmonic field in an isotropic medium is given by [7, 64]

$$\nabla^2 E_q(r,t) - \mu \frac{\partial^2 D_q(r,t)}{\partial t^2} = \mu \frac{\partial^2 P_q^{NL}(r,t)}{\partial t^2} \tag{3.9}$$

In addition, we have [7, 64]

$$D_q(r,t) = \varepsilon E_q + P_q^L \tag{3.10}$$

$$P_q^L = \varepsilon \chi_q E_q \tag{3.11}$$

where $E_q(r,t)$ is the harmonic field, $D_q(r,t)$ is the electric excitation of the harmonic field, $P_q^{NL}(r,t)$ and $P_q^L(r,t)$ are the nonlinear response and linear response of the harmonic field, respectively, and μ, ε, χ_q are the magnetic permeability, the electric permittivity and the electronic susceptibility.

Substituting equations (3.10) and (3.11) into equation (3.9), one obtains

$$\nabla^2 E_q(r,t) - \mu\varepsilon \frac{\partial^2 E_q(r,t)}{\partial t^2} - \mu\varepsilon \frac{\partial^2 \chi_q E_q(r,t)}{\partial t^2} = \mu \frac{\partial^2 P_q^{NL}(r,t)}{\partial t^2} \tag{3.12}$$

Along the laser axis (r=0), equation (3.12) may be expressed as a function of the propagation z-axis

$$\frac{\partial^2 E_q}{\partial z^2} - \mu\varepsilon \frac{\partial^2 E_q}{\partial t^2} - \mu\varepsilon \frac{\partial^2 \chi_q E_q}{\partial t^2} = \mu \frac{\partial^2 P_q^{NL}}{\partial t^2} \tag{3.13}$$

Assuming the ion is stationary, then in a non-absorbing medium the electromagnetic field of the qth harmonic along the laser axis (r = 0) for a plane wave approximation and the nonlinear response can be written as [7, 64]

$$E_q = (1/2)\left\{ \hat{E}_q \exp\left[-i(q\omega t - k_q z)\right] + cc \right\} \tag{3.14}$$

$$P_q^{NL} = (1/2)\left\{ \hat{P}_q^{NL} \exp\left[-i(q\omega t - qk_1 z)\right] + cc \right\} \tag{3.15}$$

where $\hat{E}_q$ is an envelope function (the propagation equation) of the electromagnetic field E_q and $\hat{P}_q^{NL}$ is an envelope function of the nonlinear polarization. The time t refers to the slow variation of the pulse envelope.

For the slowly varying envelope approximation, we assume [7, 64]

$$\frac{\partial^2 \chi_q E_q}{\partial t^2} \approx -q^2\omega^2\chi_q E_q \tag{3.16}$$

Moreover, the electronic susceptibility can be defined by [7, 64]

$$\chi_q = -\frac{e^2 n_e}{\varepsilon m q^2 \omega^2} \tag{3.17}$$

Substituting equations (3.16) and (3.17) into equation (3.13) gives

$$\frac{\partial^2 E_q}{\partial z^2} - \mu\varepsilon\frac{\partial^2 E_q}{\partial t^2} - \frac{\mu n_e e^2}{m}E_q = \mu\frac{\partial^2 P_q^{NL}}{\partial t^2} \tag{3.18}$$

Inserting equations (3.14) and (3.15) into equation (3.18), dividing both sides by $\exp[-i(q\omega t - qk_1 z)]$ and neglecting the double derivatives of $\hat{E}_q$ with respect to z and t in the slowly varying envelope approximation, we obtain the propagation equation of the harmonic field in an isotropic ionizing medium

$$\frac{\partial \hat{E}_q}{\partial z} + \frac{1}{v_{gq}}\frac{\partial \hat{E}_q}{\partial t} = -i\sigma\hat{E}_q + iP_q' e^{-i\Delta k_N z} \tag{3.19}$$

where $\frac{q\omega}{k_q} = v_{gq}$ is the group velocity, $k_q^2 = (q\omega)^2\mu\varepsilon$ and Δk_N is the atomic dispersion phase mismatch. $\sigma = \frac{\mu n_e e^2}{2m_e k_q}$ represents the influence of the free electrons on the wave propagation of the harmonic; $\hat{P}_q^{NL} = 2N_a(t)d_q(t)$ where $N_a(t)$ is the atomic density and $d_q(t)$ is the harmonic dipole; and $P_q^{'} = \frac{\mu q^2 \omega^2}{2k_q} \hat{P}_q^{NL}$ [64].

By making the substitutions $\eta = t - \frac{z}{v_{gq}}$ and integrating equation (3.19), the solution of the propagation equation of the harmonic field is given by

$$\hat{E}_q(\eta, z) = \frac{iP_q^{'} e^{i(-\Delta k_N/2 - \sigma/2)z} \sin[(-\Delta k_N/2 + \sigma/2)z]}{(-\Delta k_N + \sigma)/2} \tag{3.20}$$

Equation (3.20) is the same as in [75, 76]. In this equation, the effect of ionization on the harmonic propagation can be explored clearly through the parameters: Δk_N, the phase mismatch due to the atomic dispersion and σ which corresponds to the influence of the free electrons, i.e., the plasma dispersion.

Equation (3.20) indicates that the envelope of the harmonic field depends on the product of the amplitude of the harmonic dipole $d_q(t)$ (through $P_q^{'}$) and the term $\frac{\sin[(-\Delta k_N/2 + \sigma/2)z]}{(-\Delta k_N + \sigma)/2}$. Since $d_q(t) \sim I^3(t)$ [38] the time evolution of the amplitude of the atomic dipole strongly follows the laser electric field E(t). Thus, in order to investigate the influence of the ionization on the harmonic propagation, one should focus on the term $\frac{\sin[(-\Delta k_N/2 + \sigma/2)z]}{(-\Delta k_N + \sigma)/2}$.

Below, we show in the time domain calculations of the variations of the atomic dispersion phase mismatch, Δk_N, and the electron density, N_e/N_a, when the peak laser intensity is 10^{15} W/cm^2. It is useful to recall that the depletion of the neutral density scales linearly with the increase of the single ionization fraction.

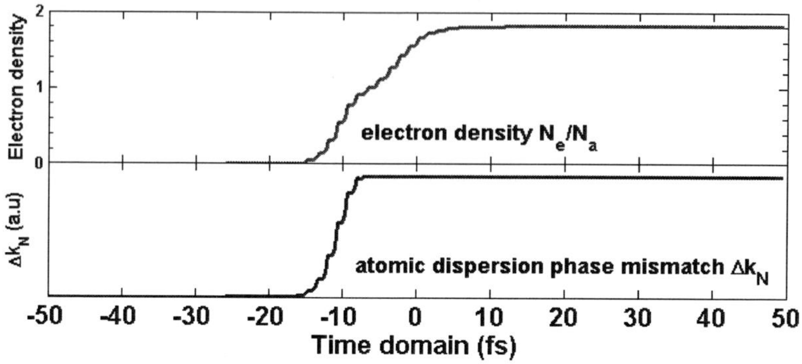

Figure 3.2. Variation of the electron density (normalized to the initial neutral density) and the atomic dispersion phase mismatch Δk_N in the time domain at $I_0=10^{15}$ W/cm^2.

Figure 3.2 indicates that for very high laser intensity (~10^{15} W/cm^2), during the ionization process, the phase mismatch which originates from the neutral gas dispersion changes much more rapidly than the free electron density which has contributions from both single and double ionizations. In addition, the term σ which indicates the influence of the free electrons on the harmonic propagation is linearly proportional to the electron density. Therefore, the variation of the neutral gas dispersion phase mismatch plays a more important role than that of σ in the study of the change of the harmonic spectral line through the term

$$\frac{\sin[(-\Delta k_N/2+\sigma/2)z]}{(-\Delta k_N+\sigma)/2}$$

[75]. However, for lower laser intensities, since the double ionization fraction is very small, the difference between the variation of the neutral gas dispersion and that of σ becomes smaller. When the double ionization is negligible (for peak laser intensities less than 4×10^{14} W/cm^2), the change of the neutral gas dispersion and that of σ tend to be the same. Under these conditions, both contribute equally to the propagation of the harmonic signal.

Figure 3.3 shows the single ionization fraction of the pulse with a peak intensity of 2.5×10^{14} W/cm^2 calculated using the Ammosov-Delone-Krainov (ADK) tunneling ionization rate. At this intensity, double ionization can be neglected.

As mentioned above, the envelope of the harmonic field depends on the product of the amplitude of the harmonic dipole, $d_q(t)$, which follows the envelope of the fundamental field, and the term

$$\frac{\sin[(-\Delta k_N/2+\sigma/2)z]}{(-\Delta k_N+\sigma)/2}.$$

Thus, during the fundamental laser pulse, the line splitting occurs only when $\frac{\sin[(-\Delta k_N/2+\sigma/2)z]}{(-\Delta k_N+\sigma)/2}$ oscillates. The bottom point between the two peaks of $\hat{E}_q(t)$ originates from the term $\frac{\sin[(-\Delta k_N/2+\sigma/2)z]}{(-\Delta k_N+\sigma)/2}$ passing through the lowest point. Consequently, a break in the harmonic envelope in the time domain causes harmonic splitting in the frequency domain [75, 76].

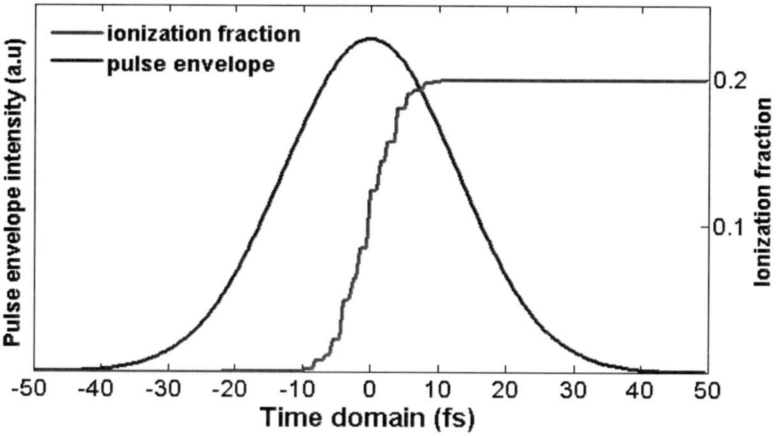

Figure 3.3. Single ionization fraction at a peak intensity of 2.5×10^{14} W/cm^2 and the laser pulse envelope in the time domain.

3.3. Experiment and Interpretation

Before we carry out experiments that allow us to study the influence of the fundamental laser intensity on the HHG spectral features, we clearly know that a modulation of the temporal phase of the fundamental laser pulses can cause a considerable change of the spectrum of the harmonics [80, 81, 82, 83]. This helps us not only to gain a better understanding of the harmonic emission process but also to have precise control over the proper harmonic spectrum required for specific applications [82, 84]. Therefore, in the first part of the experimental work, the influence of the temporal phase of the fundamental laser field on the harmonic spectrum should be investigated.

3.3.1. Influence of Phase Modulation of the Fundamental Laser Field on the Harmonic Spectrum

For a laser pulse which is centred around ω_0, it is common to expand the phase in a power series around ω_0 to second order in ω [85, 86]

$$\phi(\omega)=\phi(\omega_0)+\frac{1}{1!}\frac{\partial\phi}{\partial\omega}\bigg|_{\omega=\omega_0}(\omega-\omega_0)+\frac{1}{2!}\frac{\partial^2\phi}{\partial\omega^2}\bigg|_{\omega=\omega_0}(\omega-\omega_0)^2 \quad (3.21)$$

where $\frac{\partial\phi}{\partial\omega}\Big|_{\omega=\omega_0}$ is the first derivative of the phase at the centre frequency and corresponding to the group delay of the pulse envelope and $\frac{\partial^2\phi}{\partial\omega^2}\Big|_{\omega=\omega_0}$ is the second derivative of the phase at the centre frequency, i.e., the first derivative of the group delay, and corresponding to the group delay dispersion (GDD). Thus, the GDD which is different from zero indicates the dependence of the group delay on frequency.

To illustrate the pulse distortion due to the dispersion, it is useful to recall that the transform limited Gaussian pulse with a duration of τ_0 is given by

$$E_{in}(t)=\exp\left[-2\ln 2\left(\frac{t}{\tau_0}\right)^2\right]\exp(i\omega_0 t) \quad (3.22)$$

Using a Fourier transform, the Gaussian pulse in the frequency domain is expressed as

$$E_{in}(\omega)=\exp\left[-2\ln 2\left(\frac{\omega-\omega_0}{\Delta\omega}\right)^2\right] \quad (3.23)$$

where $\Delta\omega = 4\ln 2/\tau_0$ is the bandwidth for the Gaussian pulse.

If we consider only the influence of the GDD term, equation (3.23) can be rewritten

$$E_{out}(\omega) = \exp\left[-2\ln 2\left(\frac{\omega - \omega_0}{\Delta\omega}\right)^2\right]\exp\left[i\frac{1}{2!}\beta(\omega - \omega_0)^2\right] \quad (3.24)$$

where

$$\beta = \left.\frac{\partial^2 \phi}{\partial \omega^2}\right|_{\omega=\omega_0} \quad (3.25)$$

Using an inverse Fourier transform, one can obtain the electric field of the laser pulse after traveling through a dispersive medium in the time domain

$$E_{out}(t) = \exp(-\Gamma t^2)\exp(i\omega_0 t) \quad (3.26)$$

where $\Gamma = \left(\frac{\tau_0^2}{2\ln 2} + 2i\beta\right)^{-1}$ (3.27)

From equations (3.26) and (3.27), the effect of dispersion is disclosed. The nonlinear dependence of the phase on time which results from a complex expression of the Gaussian parameter Γ implies the existence of a time-dependent carrier frequency and this is usually understood as a chirp of the laser pulse [87]. In equation (3.27) β is defined as the frequency-sweep rate, i.e., the chirp parameter of the laser pulse. An increase of the instantaneous frequency with time ($\beta > 0$) corresponds to a positive chirp and a reverse sweep ($\beta < 0$) corresponds to a negative chirp. In other words, a positive chirped laser pulse involves a frequency increase from the leading edge to the trailing edge and a negative chirped pulse corresponds to a reverse sweep. When $\beta = 0$, the frequency of the laser field exhibits an oscillatory periodicity and this is understood as a chirp-free or transform-limited condition.

In addition, from equation (3.27), one finds that the pulse duration depends on the chirp parameter β and is given by

$$\tau^2 = \tau_0^2 + \left(\frac{4\beta \ln 2}{\tau_0}\right)^2 \qquad (3.28)$$

If without increasing the laser energy, to keep the peak intensity fixed for the longer pulse durations, based on equation (3.28), for positive chirped pulses ($\beta > 0$) or for negative chirped pulses ($\beta < 0$), the pulse is always stretched leading to a reduction of the peak laser intensity. This means that at the chirp-free condition, the pulse duration is shortest and the peak laser intensity is highest.

In a chirped-pulse amplification (CPA) laser, the chirp parameter, i.e., the phase of the pulse, can be varied by adjusting the grating separation in the pulse compressor.

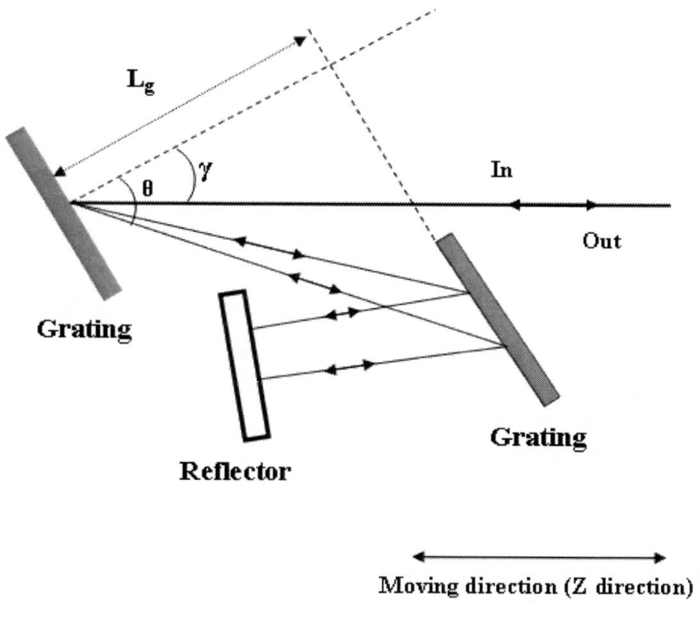

Figure 3.4. Pulse compressor based on a grating pair.

The pulse compressor of a CPA consists of two parallel gratings and a reflector as illustrated in figure 3.4. This grating pair is installed not only to compress a laser amplified pulse to achieve high output intensity but also to compensate the dispersion induced by the stretcher and all the other components of the CPA laser system by choosing the proper distance between these two gratings.

The group delay dispersion which is produced by the compressor can be calculated from [88]

$$\mathrm{GDD} = \frac{\partial^2 \phi(\omega)}{\partial \omega^2} = -\frac{\lambda^3 L_g}{\pi c^2 d^2}\left(1-\left(\frac{\lambda}{d}-\sin\gamma\right)^2\right)^{-3/2} \tag{3.29}$$

where d is the groove separation, γ is the angle of incidence on the first grating and L_g is the distance between the two gratings.

From equation (3.29), the phase of the fundamental laser field is modified when the distance between the two gratings is varied along the incident direction of the laser beam to keep the alignment. When the grating separation is changed by δZ along the incident direction of the laser beam, the change of L_g is given by [88, 89]

$$\delta L_g = 2(\cos\gamma / \cos\theta_0)\delta Z \tag{3.30}$$

where $\theta_0 = \theta(\omega_0)$ is the diffraction angle corresponding to the central frequency.

Provided the compressor fully compensates for the dispersion at the grating separation Z_0 along the laser incident direction, the change of phase is determined by the grating detuning which is given by $\delta Z = Z - Z_0$. Thus, a zero grating detuning occurs when $Z = Z_0$ corresponding to complete compensation for the dispersion. Consequently, for a zero grating detuning, the pulse duration is shortest. Based on the linear relation between the grating detuning δZ and δL_g shown in equation (3.30) and that between the group delay dispersion and the distance between these two gratings shown in equation (3.29), it is clear that when the grating separation is changed by δZ along the incident direction of the laser beam from the zero grating detuning, the chirp parameter β is given by [84]

$$\beta = -C \times \delta Z \tag{3.31}$$

where C is a positive constant.

Equation (3.31) indicates that a reduction of the grating separation in the pulse compressor from the chirp free condition, i.e., a grating position with negative detuning $\delta Z < 0$, causes a positive chirp while an increase of the grating separation causes a negative chirp.

In the experimental work, first, the laser pulse is set at the transform-limited condition (chirp free) and therefore the shortest pulse duration of ~ 30 fs is obtained. After that, experimental parameters such as the argon gas pressure, the aperture diameter and the position of the laser focus are controlled so that all available harmonics appear intense for the on-axis emission and spectrally narrow which is typical for the dominant contribution from the short trajectory emission. Then, the chirp of the laser is varied from a negative sign to a positive sign by adjusting the separation of the gratings. The pulse duration at the largest positive chirp and the largest negative chirp is estimated to be 200 fs. From equation (3.28) and based on the shortest pulse duration at the transform-limited condition (τ_0 ~ 30 fs) and the measured pulse duration at the largest positive chirp and largest negative chirp $(\tau \sim 200 fs)$, we can calculate the chirp parameter, i.e, the frequency-sweep rate, β, as shown in figure 3.5 which plots the harmonic spectrum generated from argon gas as a function of β. It is useful to recall here that theoretically the chirp parameter scales linearly with the grating separation while the pulse duration scales quadratically with this separation.

Figure 3.5 indicates that a variation of the chirp of the laser pulse can alter the harmonic spectrum. Firstly, the cut-off energy of the observed spectrum is shifted when the chirp is adjusted. The highest cut-off energy is about 48 eV (H31) and it is reduced to 44.7 eV (H29), 41.7 eV (H27), 38.6 eV (H25) and 35.6 eV (H23) in succession when the chirp is moved away from zero (toward to a larger chirp on the positive or negative sides). This may be attributed to a reduction of the peak intensity of the fundamental laser which occurs when the pulse is stretched. Indeed, when varying the chirp from negative to positive, H23 appears for a pulse duration of ~ 110 fs, H27 appears for a pulse duration of ~ 80 fs and H31 appears for a pulse duration of ~ 40 fs. The decrease of the peak laser intensity also causes a rapid reduction of the harmonic intensity when the chirp is varied from the transform-limited condition.

Also, there is a spectral shift of the harmonics and this spectral shift for all harmonics follows the same trend. Clearly, in figure 3.5, the harmonics produced by large chirped pulses on both the positive and negative sides are red-shifted relative to those produced by free chirped pulses and the shift on the positive side is much stronger than that on the negative side. Figure 3.6 shows the measured maximum energy shift ($\Delta E_{q,max}$) for different harmonics.

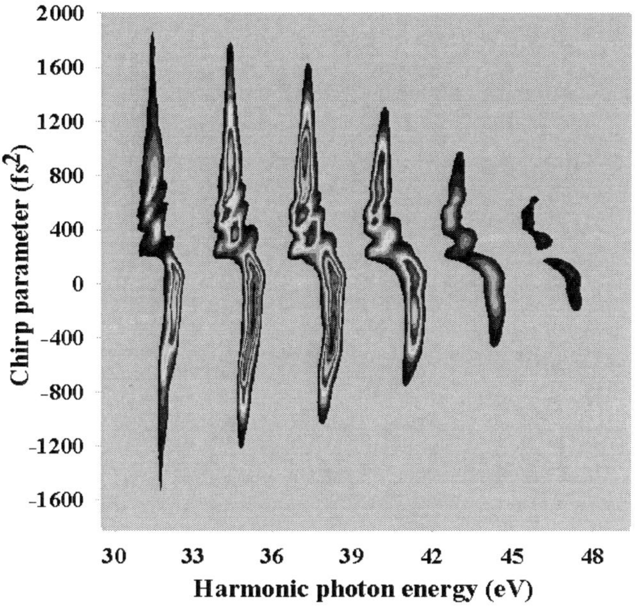

Figure 3.5.Variation of the harmonic spectrum as a function of the chirp parameter, β.

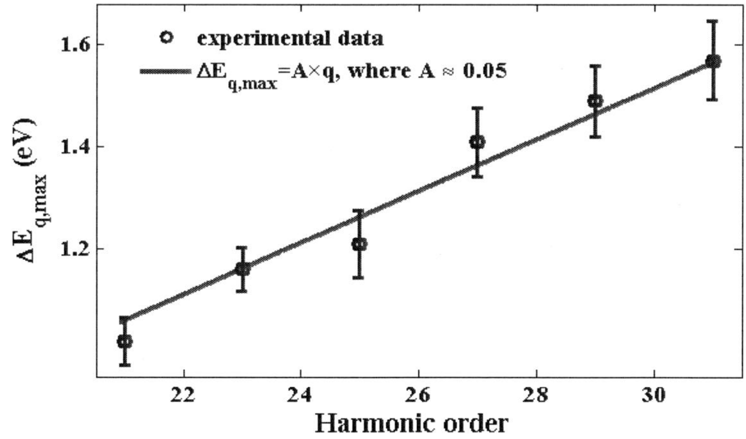

Figure 3.6.Maximum energy shift for different harmonics.

From figure 3.6, we find that the maximum energy shift scales almost linearly with the harmonic order ($\Delta E_{q,max} \sim q$). This implies that the frequency shift of the fundamental laser field leads to the shift of the harmonic spectrum, $\Delta\omega_q = \Delta\omega / q$.

In order to be more apparent for further analysis, figure 3.7 plots the spectrum of H25 which has the highest intensity among the available harmonics when the fundamental laser pulse is transform-limited as a function of the chirp parameter β.

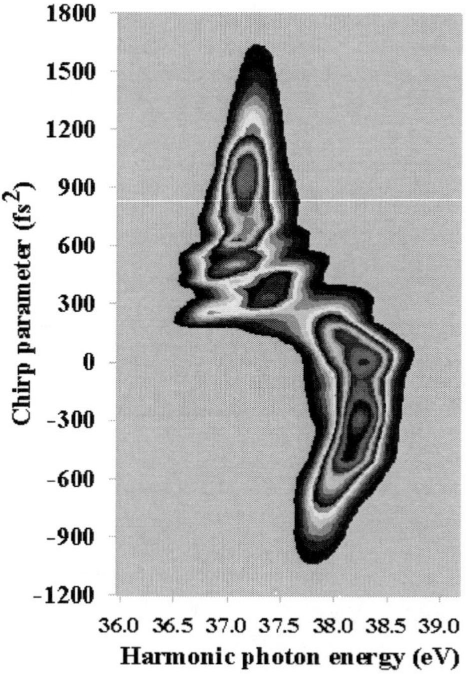

Figure 3.7.Variation of the H25 spectrum as a function of the chirp.

In figure 3.7, the spectrum at the transform-limited condition exhibits a redshift for both positive and negative chirps in which the redshift on the positive sign ($\Delta E_{max,positive}$ ~ 1.21 eV) is much more pronounced than that on the negative sign ($\Delta E_{max,negative}$ ~ 0.3 eV). That the harmonic peak strongly shifts to longer wavelengths for positive chirp is consistent with the observation in [9, 80]. As stated above, this behaviour can be easily explained in terms of the frequency change of the fundamental field on the leading edge of the laser pulse when the pulse is chirped. On the other hand, with negatively chirped pulses the harmonic spectrum is well defined and exhibits a slight redshift. This behaviour which is different to that observed in [9, 80] indicates that in our experiment the positive chirp of the fundamental laser field induced by the dispersion associated with the propagation of the laser field in the ionized gas medium and through the focusing lens strongly affects the spectral structure of the harmonic. This positive chirp is compensated by the negative chirp of the laser pulse leading to a sharp and intense harmonic spectrum with the maximum intensity occuring for a negative chirp ~ 200 fs^2 (~ 35 fs in pulse duration). On the other hand, this positive chirp can be added to the positive chirp of the laser pulses resulting in a strong redshift and a broad spectrum which is seen in figure 3.7. We should note that the positive chirp induced by the propagation of the laser field in the ionized medium depends on the laser intensity which is strongest when the pulse duration is shortest; therefore the influence of this positive chirp on the harmonic spectrum becomes weaker when the pulse duration is stretched. This may explain why in figure 3.7, for the positively chirped pulses, the harmonic spectrum exhibits a significant redshift for a chirp less than 500 fs^2 (~ 55 fs in pulse duration). In addition, the strong influence of two different positive chirps (the positive chirp induced by the propagation of the fundamental laser field on the ionized medium and through the focusing lens when β is between 250 fs^2 to 450 fs^2 and the positive chirp induced by the variation of the grating separation in the pulse compressor) may also cause the considerable modulation of the spectral line which is observed in figure 3.7 in this region of β. For larger positive chirps, the spectral structure exhibits a narrow line without any shift and modulation and a sharp decrease in intensity due to the reduction of the fundamental laser intensity.

We are also aware that the observed weak spectral redshift on the negative chirp side in figure 3.7 may be mostly associated with the slight redshift of the fundamental laser pulse that occurs when the laser intensity rapidly decreases. Indeed, for large negative pulses, i.e., a large increase of the pulse duration, since the laser intensity is very weak, the effect of the positive chirp induced by the dispersion of the ionized medium on the modulation of the spectral feature becomes minor and can be neglected.

In summary, we have shown the influence of the laser chirp on the spectral structure of the harmonics. In the experiment, variation of the chirp not only reduces the peak laser intensity due to the longer pulse duration but also leads to a temporal phase modulation of the fundamental field. Thus, the harmonic spectrum dramatically changes with the chirp of the fundamental laser pulses. First, the maximum cut-off energy is shifted from 48 eV down to 35.6 eV when the chirp is moved away from the transform-limited condition because of the rapid decrease of the peak laser intensity. Secondly, the spectral shifts of the harmonics are mainly due to the frequency shift of the fundamental field and we find that the maximum energy shift is 1.58 eV for the highest harmonic order H31 when the chirp is varied from the transform-limited condition.

3.3.2. Influence of the Driving Laser Intensity on Spectral Feature of HHG

In the previous section, we demonstrated that the phase of the fundamental field strongly affects the HHG process. On the other hand, as presented in the theoretical background, the harmonic yield and harmonic spectrum are also considerably influenced by the fundamental laser intensity through the macroscopic phase mismatch and the atomic dipole. Thus, in this section, the experimental work involving the fundamental laser intensity dependence of the HHG process is presented.

According to the theoretical studies in sections 3.2.2 and 3.2.3, one knows that during the laser pulse phase matching is transiently optimized at certain intensities I_n corresponding to a given time t_n and can be expressed in terms of the peak intensity I_0 as shown in equation (3.7). Moreover, the ionization fraction also strongly affects the HHG process. Hence, for different initial values of the laser energy which are applied to produce the harmonic radiation and limitthe ionization rate, the behaviour of the harmonic yield and the harmonic spectrum as a function of the laser energy may be different. In this work, the changes of the harmonic yield and the harmonic spectrum with variation of the laser energy are experimentally analyzed in two cases: without limitation of ionization and with limitation of ionization.

3.3.2.1. Experimental Setup

The laser system was reported in section 1.2.1. Pulses with a duration of 30 fs centred at 805 nm are focused by a 300 mm focal length lens into a 150 mm long gas cell with a glass window at the entrance and a 100 μm pinhole at the exit. A half-wave plate and a polarizing beam splitter are used to control the laser energy precisely and continuously without any change of the focusing geometry. Argon gas is used. The harmonic production and detection systems are the same as those in section 1.2.2.

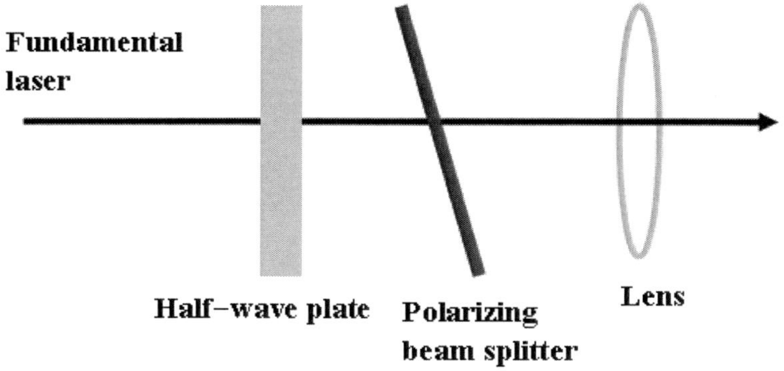

Figure 3.8.A combination of a half-wave plate and a polarizing beam splitter to control the laser energy.

3.3.2.2. Study of Driving Laser Intensity Dependence of the Harmonic Spectral Features without Limitationof Ionization Fraction

First, the half-wave plate and the polarizing beam splitter are positioned so that the laser energy with full beam coming after the polarizing beam splitter is very low $(\sim 0.4\,\mathrm{mJ})$. In the following, the laser focus is set close to the exit of the gas cell and other experimental parameters such as the gas pressure and the aperture diameter are controlled so that a maximal photon flux and a narrow spectrum are observed for all available harmonics. Under these conditions, the energy of the fundamental laser is measured at ~ 0.4 mJ which corresponds to a very large diameter of the aperture (~ 10 mm diameter), i.e., the aperture is almost fully opened for the laser beam. Later, all the experimental parameters (e.g., the gas pressure and aperture diameter) are kept unchanged and only the laser pulse energy is varied continuously from ~ 0.2 mJ, at which the harmonics start to be observed, to a maximum value of 3.6 mJ by rotating the half-wave plate. Thus, in this case the ionization fraction can be much higher than the critical value when the laser energy increases.

Figure 3.9 shows the harmonic spectrum at a laser energy of 0.4 mJ using argon gas.

Figure 3.10 shows the H21 intensity, which is the strongest harmonic in the HHG spectrum shown in figure 3.9, as a function of the laser energy. The maximum photon flux of H21 during variation of the laser energy is $\sim 1.5\times 10^{10}$ photon $/(\mathrm{cm}^2\mathrm{s})$.

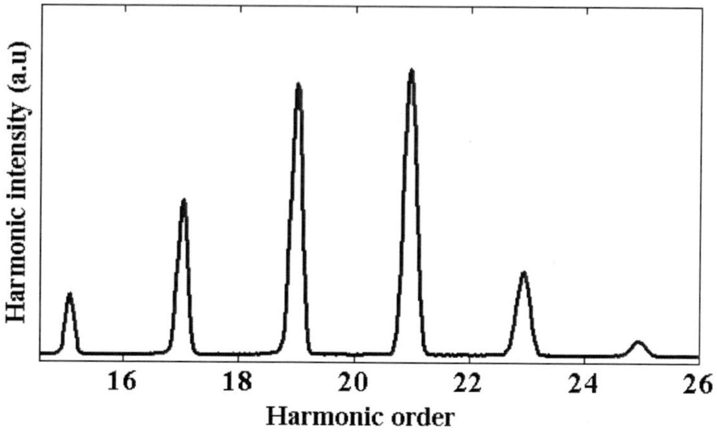

Figure 3.9.HHG spectrum with argon gas when the laser energy is 0.4 mJ.

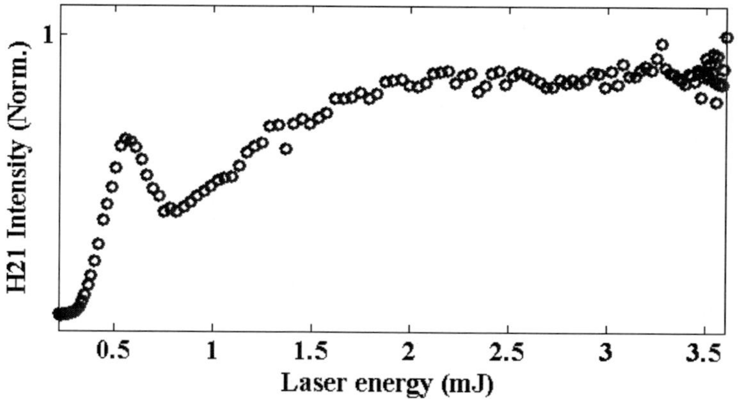

Figure 3.10.Intensity of harmonic H21 for argon when varying the laser energy of the fundamental laser.

In figure 3.10, the development of the harmonic intensity with laser energy, i.e., the harmonic conversion efficiency, is relatively complicated and slow which may be explained by a large plasma phase mismatch resulting from the very high free electron density in the medium. Moreover, the defocusing effect due to the high ionization fraction reduces the focusing. Consequently, a saturation of the harmonic intensity can be seen for laser energies higher than 2.3 mJ.

As shown theoretically in sections 3.2.2 and 3.2.3, for high laser energy, the influence of the modulation of the harmonic dipole phase and the high ionization on the harmonic spectrum can be clearly observed. In addition, the change of the harmonic spectral line for the off-axis region which has the dominant contribution of the long trajectory with large coefficient, α_q, would be stronger than that for the on-axis region. Therefore, firstly, we measure the harmonic spectrum of the H21 off-axis selection as a function of the laser energy.

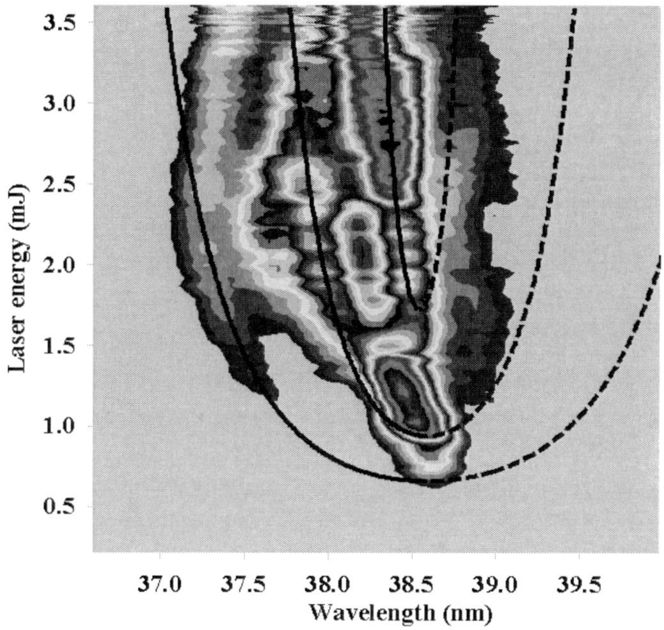

Figure 3.11.Measured H21 spectrum of the off-axis selection as a function of the laser energy. The black lines are calculated from equation (3.32), fitted to the experimental spectrum.

In figure 3.11, spectral interference fringes are visible. This is similar to the results obtained in [71] and may also be attributed to transient phase matching during the pulse. A first fringe with a maximum blueshifted wavelength of ~ 1.4 nm, a second fringe with a maximum blueshifted wavelength of ~ 0.8 nm and a third fringe with a maximum blueshifted wavelength of ~ 0.2 nm are observed for laser energies ~ 0.6 mJ, 0.9 mJ and 1.6 mJ, respectively.

Although the strongly ionized medium may significantly influence the spectral properties of the fundamental laser, this only causes an overall spectral shift, not double peak structures of the driving laser field [66]. This confirms that the spectral features in figure 3.11 do not originate from the structure of the fundamental spectrum.

The blueshifting and splitting observed in the experiment may be associated with the strong harmonic phase modulation which is indicated by equation (3.8). To compare the experiment with the theory, we perform a calculation of the H21 spectrum as a function of the laser pulse energy. Based on equation (3.8), one can obtain an equation which describes the exponential shape of the maximum signal in the harmonic spectrum plotted as a function of the laser intensity [71]

$$I_0(\omega_q) = I_n \exp\left[\frac{\sigma^2(q\omega - \omega_q)^2}{4\alpha_q^2 I_n^2}\right] \tag{3.32}$$

In this calculation, we estimate that the diameter of the focus point is around 150 μm. For simplification, we neglect the defocusing effect in this laser energy range and then the laser intensity axis can be converted linearly from the laser energy axis. For different intensities I_n, i.e., different fringe patterns, the coefficient, α, is adjusted to obtain the best agreement with the experimental harmonic spectral lines. The simulated results are shown by the black lines in figure 3.11. We find that the maximum coefficient is around 11.5×10^{14} cm^2/W which is consistent with a larger contribution from the long trajectory in the off-axis region.

As shown by the simulations, the exponential shape of the harmonic spectral distribution may be visible. However, in the experiment, we cannot see clearly the red wings (the dashed lines) and this is likely to be due to the significant plasma phase mismatch induced by the strong ionization rate on the trailing edge of the pulse which distorts the harmonic emission on this edge. This behaviour also allows us to confirm that the quantum path interference (QPI) is not principally responsible for the interference fringes observed in our experiment. For QPI, as indicated in [72], changes in phase matching may also occur with increased intensity but the shapes with both the blue and red wings still appear at low intensities.

Unlike the simulations, the experimental spectral lines do not shift anymore for laser energies higher than 2.3 mJ. As explained above, this is a direct consequence of the almost saturation of the laser intensity induced by the defocusing effect and this is also seen through the saturation of the harmonic intensity for laser energies higher than 2.3 mJ in figure 3.10. We note that in practice, in the case without limitation of ionization, the defocusing effect can even occur for laser energies less than 2.3 mJ. This is one of reasons for the difference between the simulated and measured harmonic spectrum in this laser energy regime.

We have already studied the spectrum of the harmonic emission in the off-axis region as a function of the laser energy. Now, we investigate the variation of the spectral line in the on-axis region where theoretically the short trajectory predominantly contributes to the harmonic radiation.

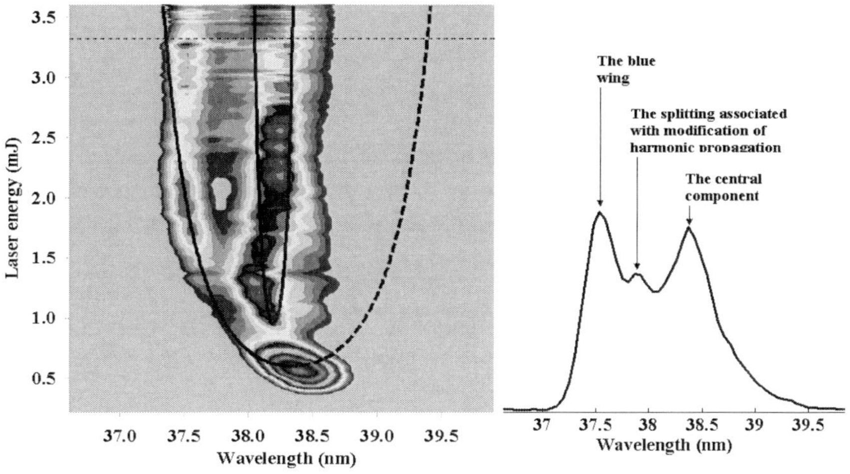

Figure 3.12.Measured H21 on-axis selection as a function of the laser energy. The black lines are calculated from equation (3.32), fitted to the experimental spectrum (a); and the harmonic spectrum for a laser energy ~ 3.3 mJ (b).

In figure 3.12, for laser energies less than 0.6 mJ, as the laser energy increases, the harmonic signal increases and the spectrum still appears to be narrow. For laser energies greater than 0.6 mJ, the spectrum exhibits a blueshifting causing the first fringe. We would observe a second fringe when the laser energy is ~ 0.9 mJ. However, because this shifting may be very small a second fringe cannot be seen. In this case, we just see the central component.

Similarly to the case of the off-axis selection, by performing a simulation and fitting to the experimental results, we find a maximum coefficient ~ 7×10^{14} cm^2/W which is in agreement with the major contribution of the short trajectory. However, this value is higher than the theoretical values given by [73] which indicates a non-negligible contribution of the long trajectory for the harmonic emission on the spectral axis. Apart from the defocusing effect, the simultaneous contributions of the long and short trajectories to the harmonic generation may lead to the fact that the measured harmonic spectrum becomes more complicated than the calculated ones.

Similarly to figure 3.11, the spectral shape in figure 3.12 is also asymmetric, i.e., the red wing of the first fringe (the dashed line) is not visible. This phenomenon originates from the distortion of the harmonic emission from the trailing edge due to the significant plasma phase mismatch. Furthermore, the observation that the experimental blueshifted wavelength achieves a maximum value (~ 0.9 nm) when a pulse energy of 2.3 mJ is applied is also understood by the almost saturation of the laser intensity induced by the defocusing effect.

Interestingly, when the laser energy is above 3 mJ, we can observe a slight splitting between the blue wing of the first fringe and the central component. The splitting becomes stronger with increase of the laser energy and may be attributed to the modification of the harmonic propagation in the ionizing medium as mentioned theoretically in section 3.2.3. Indeed, without limitation of the ionization fraction, the rapid ionization of the gas can strongly modify the harmonic propagation and this also leads to a more complicated experimental harmonic spectrum compared to the calculated ones based on equation (3.32). In order to see this splitting clearly, figure 3.12b shows the measured harmonic spectrum for a laser energy ~ 3.3 mJ.

Based on the harmonic spectral features shown above, it appears that when the aperture is almost fully open for the laser beam during the scanning of the laser energy (similar to without an aperture), there is no limitation of the ionization fraction at a high specific laser energy and there is no control of the wavefront of the driving pulse. Thus, we are not able to produce a sharp and strong harmonic spectrum during the scanning of the laser energy from minimum to maximum value. In this case, with an increase of the laser energy the harmonic conversion efficiency is very low due to significant plasma phase mismatch which can be seen in figure 3.10 and the harmonic spectrum exhibits strong shifting and splitting which can be seen in figures 3.11 and 3.12.

3.3.2.3. Study of Driving Laser Intensity Dependence of the Harmonic Spectral Features with Limitation of Ionization

In this study, the experimental setup is the same as in figure 3.8. However, at the beginning, the half-wave plate and the polarizing beam splitter are positioned so that the laser energy coming after the polarizing beam splitter is maximal (~ 4 mJ). The laser focus is set close to the exit of the gas cell. Other experimental parameters such as the aperture diameter and the gas pressure are controlled so that all available harmonics appear to have intense on-axis emission and to be spectrally narrow which is typical for a dominant contribution from the short trajectory emission. Under these conditions, the energy of the fundamental laser is measured to be 2.1 mJ and corresponding to a reduced diameter of the aperture of the full beam (~ 6 mm in diameter). Later, all of the experimental parameters (e.g., gas pressure and aperture diameter) are kept unchanged and the laser pulse energy is varied continuously from a minimum value of 0.3 mJ, at which point the harmonics start to be observed, up to a maximum value of 2.1 mJ by rotating the half-wave plate. The gas pressure is kept relatively low (~ 70 Torr) to minimize absorption effects. By using the aperture to control the ionization fraction at the maximum laser energy, we can ensure that during scanning of the laser energy, the ionization fraction is kept below the critical ionization level (e.g., for argon, the critical ionization fraction is less than 5 % [34]). Figure 3.13 shows the observed harmonics at a laser energy of 2.1 mJ using argon.

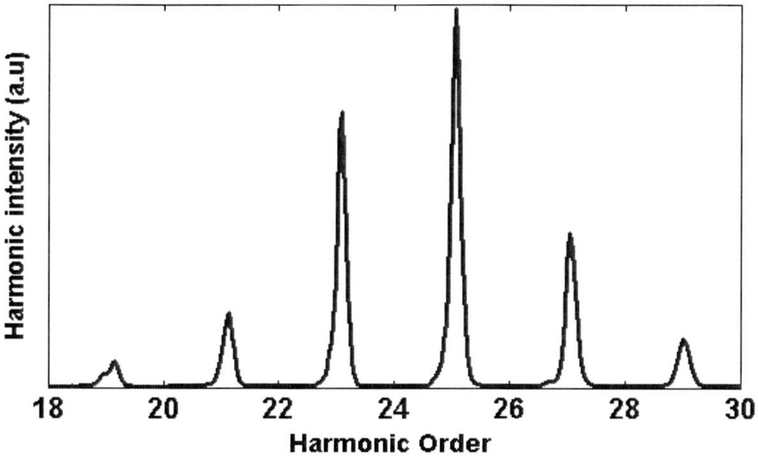

Figure 3.13.HHG spectrum with argon gas when the laser energy is 2.1 mJ.

Figure 3.14 shows the spectrum of the four strongest harmonics (figure 3.14a) and the selected harmonic H25 (figure 3.14b) as a function of the laser energy without any broadening or splitting.

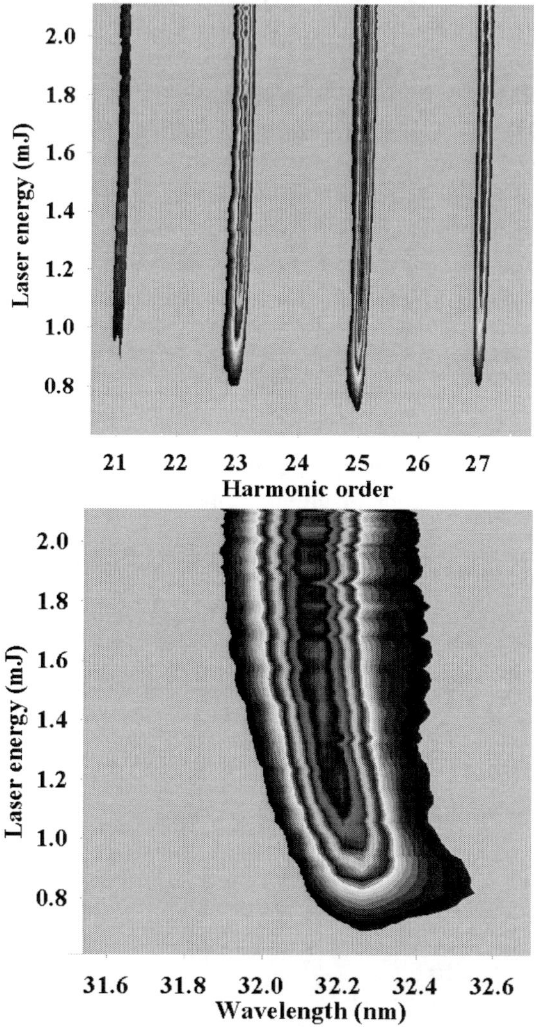

Figure 3.14. Measured harmonic spectrum (a) and selected harmonic H25 (b) as a function of the laser energy.

The spectrum remains narrow with high intensity on-axis during scanning of the laser energy from the minimum to the maximum value. This behaviour indicates that the short trajectory whose coefficient, α_q, is relatively small contributes predominantly to the harmonic emission during the scanning of the laser energy. In addition, the influence of the ionization and the harmonic dipole phase, which vary strongly during scanning of the laser energy, on the

HHG process cannot be observed by examining the harmonic spectrum in this case. This suggests that these influences may be revealed by investigating the laser intensity dependence of the harmonic yield.

In order to perform this task, we measure the intensity of the on-axis harmonic H25 when the laser energy varies from 0.6 mJ to 2.1 mJ (the H25 harmonic appears when the laser energy is around 0.6 mJ). The maximum photon flux of H25 during variation of the laser energy is $\sim 9\times10^{11}$ photon/(cm^{2}s).

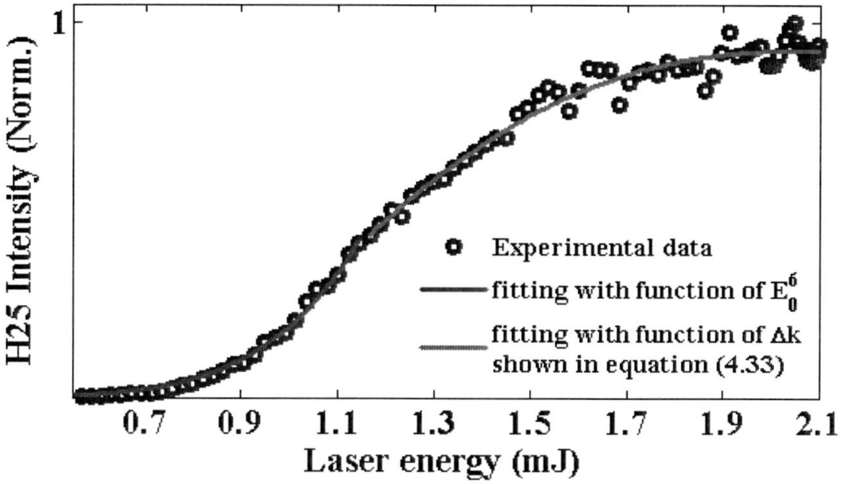

Figure 3.15.Intensity of the harmonic H25for argon gas when varying the laser energy of the fundamental laser.

In figure 3.15, it is clear that for low laser energies (below 1.2 mJ), the H25 intensity varies rapidly with the laser energy. The development of the harmonic signal in this region can be fitted very well with the function E_0^6, where E_0 is the laser energy. Indeed, for low laser energies, the ionization fraction is very small, i.e., the defocusing effect can be ignored and therefore one can assume that the diameter of the focus point remains unchanged. Consequently, in this region, the peak laser intensity, I_0, is directly proportional to the laser energy, E_0, and this leads to the fact that the harmonic yield can be also fitted by the function I_0^6. In addition, for low laser energies, there is a minor influence of the ionization on the phase mismatch and the variation of the atomic dipole phase is also negligible since it scales linearly with the laser intensity. This means that macroscopic effects can be neglected and only the single-atom response is exhibited in this regime.

When the laser energy is above 1.2 mJ the ionization fraction becomes higher and, especially for the gas cell geometry, the rapid increase of ionization causes strong plasma dispersion and depletion of the neutral gas density. The laser intensity dependence of the harmonic yield in this region becomes more complicated. Indeed, in figure 3.15, for $E_0 > 1.2$ mJ, it is shown that the development of the harmonic intensity with laser energy, i.e., the harmonic conversion efficiency, is slower due to the strong influence of the phase mismatch. In addition, the defocusing effect due to the high ionization fraction causes reduced focusing. Therefore, the laser intensity does not scale linearly with the laser energy anymore. The considerable phase mismatch and the defocusing leads to almost saturation of the harmonic yield at a laser energy of $\sim 1.9\,\mathrm{mJ}$.

In the high energy regime, the development of the harmonic intensity as a function of E_0 can be fitted by the equation

$$I_q = a_0\left(1 - a_1.\frac{\sin^2\left(\dfrac{\Delta k_{q,\mathrm{eff}} L_{\mathrm{med}}}{2}\right)}{\left(\dfrac{\Delta k_{q,\mathrm{eff}} L_{\mathrm{med}}}{2}\right)^2}\right) \tag{3.33}$$

where $\dfrac{\Delta k_{q,\mathrm{eff}} L_{\mathrm{med}}}{2} = a_2 + a_3 E_0$ (3.34)

and $\Delta k_{q,\mathrm{eff}}$ is an effective phase mismatch which is proposed in order to simplify our calculation.

The dependence shown in equation (3.33) suggests that when E_0 is higher than 1.2 mJ, the phase mismatch plays a significant role in the variation of the harmonic yield and results in a slow rise of the harmonic conversion efficiency as plotted in figure 3.15. Moreover, with a constant a_0 in the fitting equation, it is shown that the amplitude of the harmonic dipole moment remains nearly unchanged. The linear dependence of the phase mismatch on laser energy indicated in equation (3.34) reflects the more important contribution of the harmonic dipole phase term to the phase mismatch than the plasma dispersionand the neutral gas dispersion. This makes sense since we initially keep the ionization fraction not higher than the critical level at the maximum laser energy.

In figure 3.14b, for laser energies less than ~ 1.8 mJ, we observe a small blueshift ($\Delta\lambda < 0.1$ nm) in the H25 spectrum with increasing laser energy. This may be due to the larger contribution of the electrons accelerated in the leading edge of the laser pulse to the harmonic generation process and the dominant harmonic emission with the short trajectory. However, for laser energies higher than ~ 1.8 mJ, the defocusing effect causes almost saturation of the laser intensity and therefore the shift is no longer visible.

Based on figures 3.14 and 3.15, we also find that when an aperture is used to control the ionization fraction at a high specific laser energy less than the critical ionization and to control the wavefront of the fundamental laser beam, we are able to generate strong and sharp harmonics with a dominant contribution from the short trajectory.

CONCLUSION

We have presented the influence of the driving laser on the spectral features of HHG. It has been shown that not only the temporal phase but also the intensity of the fundamental beam strongly affects the HHG process. Indeed, the harmonic spectra involving the cut-off energy and the spectral shape dramatically changed with the variation of the laser chirp. The maximum cut-off energy was shifted from 48 eV down to 35.6 eV and the maximum energy shift was 1.58 eV for the highest harmonic order H31 when the chirp was moved away from the transform-limited condition. In addition, in this chapter through studies of the laser intensity dependent harmonic generation, the influence of the harmonic dipole phase and the dispersion phase mismatches induced by a strongly ionized gas medium on the harmonic spectrum were clearly revealed. When there was no limitation of ionization fraction, spectral interference fringes in which a first fringe with a maximum blueshifted wavelength of ~ 1.4 nm, a second fringe with a maximum blueshifted wavelength of ~ 0.8 nm and a third fringe with a maximum blueshifted wavelength of ~ 0.2 nm have been observed for laser energies ~0.6mJ, 0.9 mJ and 1.6 mJ, respectively. Furthermore, we have shown that use of an aperture to limit the ionization fraction to less than a critical level (~ 5 % in argon) at a high specific laser energy (~ 2.1 mJ) and to control the wavefront of the driving laser pulse for strongly coherent buildup of the on-axis harmonic radiation is critical for generating a sharp and intense harmonic spectrum.

Chapter 4

PHASE MATCHED HIGH HARMONIC GENERATION FOR THE STUDY OF ALIGNED MOLECULES

4.1. INTRODUCTION

HHG is produced not only from atoms but also from molecules. Like HHG from atoms, HHG from molecules can also be understood in term of a three-step model as mentioned in section 1.1.2.2. Studies of the process of high harmonic generation can provide an understanding of the microscopic and macroscopic processes and may lead to information about the structure and dynamics of the atom or molecule.

Unlike atoms, molecules and especially linear diatomic molecules such as N_2 or O_2 are not isotropic systems. For randomly aligned molecules, the HHG has been shown to have characteristics similar to those from atoms, but for aligned molecules the HHG is influenced by the angle between the molecular axis and the polarization vector of the femtosecond laser field [90]. Alignment of molecules has been used to study molecular orbital reconstruction [91] quantum interference [92], nonlinear frequency conversion [93] and filamentation control [94]. On the experimental side, it is interesting to study the HHG yield for field-free aligned molecules produced by the interaction of the molecules with a short, relatively weak pulse to create a rotational wave packet. The revival of the dephased wave packet at intervals separated by their fundamental rotational period leads to the periodic alignment of the molecules. In order to observe the alignment dependence of the HHG, a delayed higher intensity laser pulse is used to generate HHG at different time intervals after the pre-aligning pulse when the molecules undergo a rapid change in their alignment. By varying the delay time between the pulses the HHG emission intensity for different magnitudes and angles of alignment can be detected and used to study the rotational molecular dynamics [90-92, 95-97]. However, not only the ionization and electron-ion recombination but also the phase matching condition depends on the alignment of the molecules and this does not appear to have been considered in previous studies.

The phase-matched propagation of the fundamental and harmonic radiation, i.e., the coherent construction of the XUV field in the macroscopic medium, is reflected in the observed harmonic intensity. As indicated in section 1.1.3.2, the degree of phase-matching depends on the harmonic order and on a number of experimental parameters including the geometry, the atomic and molecular dispersion, the absorption coefficient of the target gas at the harmonic frequencies, the ionization fraction, and the gradient of the atomic or molecular dipole phase. Some of these parameters depend on the molecular alignment and the molecular structure. Therefore, both the microscopic single-molecule response to the laser field and the macroscopic phase-matching of the harmonics and the laser field will play an important role in determining the alignment dependence of the harmonic radiation from molecules, and this needs to be considered, especially when a high gas density or a long interaction length is used.However, HHG is a complex high-order nonlinear process, and only the microscopic single-molecule response has been considered in previous studies [90-92, 95-97] which may lead to contradictory results when studying molecules with a complicated ground state, such as N_2.

On the other hand, for linear molecules excited by a short laser pulse in free space, a modulation of the refractive index in space and time is induced along the direction of the field polarization corresponding to various molecular alignment revivals. Based on this, several types of rotational coherence spectroscopy (RCS) have been developed on a picosecond or femtosecond time-scale in order to obtain structural and dynamical information on molecular gases [98, 99]. Since the coherent rotational states prepared in a RCS experiment experience a dephasing due to intra- and inter-molecular processes, such experiments, in addition to yielding spectroscopic constants, are able to measure the coherence relaxation time. However, the technique is difficult to apply to molecules with different symmetry of the ground-state configuration and requires a low temperature molecular beam [100, 101].

In this chapter, we discuss the important role of phase-matching on the study of high harmonic generation from aligned molecules and propose a new approach to obtain information about molecular structure dynamics based on phase matching harmonic generation in a semi-infinitive gas cell.

4.2. ALIGNMENT OF MOLECULES IN INTENSE LASER FIELDS

When an intense polarized laser beam interacts with molecules, there is an alignment of the molecules. This is the process in which one or more molecular axes become confined along the fixed axes determined by the polarization state of the laser beam. There are two kinds of alignment: an adiabatic alignment and a non-adiabatic alignment. In terms of the physical mechanism, adiabatic alignment occurs when the laser is turned on and off slowly compared to the rotational period of the molecule. When there is an interaction with an external laser field, the molecules orient themselves in the direction of the laser field. When the laser pulse fades away, the molecular ensemble returns smoothly to an isotropic angular distribution. On the other hand, non-adiabatic alignment requires a more quantum mechanical description for a molecular ensemble and occurs when the duration of the laser pulse is much shorter than the rotational period of the molecule. In this case, the interaction with an external laser field leaves the molecule in a coherent superposition of rotational eigenstates. A rotational wave packet freely evolving in time suffers a periodic dephasing and rephasing of its components; thus the aligned state only lasts for a short while, but it is periodically reconstructed at multiple revival times which persist for a long time after the interaction. In our studies presented in this chapter, the latter process is implemented.

4.2.1. Adiabatic Alignment

In this alignment, the optical field pulse duration is much longer than the rotational period of the molecules. In this adiabatic limit, the unaligned molecules are slowly oriented as the laser is slowly turned on and the degree of alignment reaches its maximum value at the peak of the laser pulse. When the laser is slowly turned off, the alignment gradually decreases and finally disappears completely. In other words, the degree of alignment follows the instantaneous laser intensity and therefore, for a given molecule, it depends on the duration, the intensity and the wavelength of the laser beam [99, 102].

In the observation of adiabatic alignment, one dimensional alignment can be obtained by applying a linearly polarized laser beam. In this case, the most polarizable axis of a molecule is confined along the polarization axis of the linearly polarized laser beam.Two and three dimensional alignments can be observed with circularly and elliptically polarized fields, respectively.

Adiabatic alignment has some advantages. For example, it is easy to experimentally implement and the alignment following the laser intensity profile lasts as long as the laser field is applied. However, when using an intense laser beam, since the duration of the strong field is so long, it can destroy processes and the mechanism of some applications. Therefore, it is necessary to propose another alignment technique which is non-adiabatic alignment.

4.2.2. Non-adiabatic Alignment

Opposite to adiabatic alignment, non-adiabatic alignment is obtained when the optical field pulse duration is much shorter than the rotational period. This alignment is also known as field-free alignment because the presence of the laser field is very short.

In a classical model, the electric field $\vec{E}(t)$ induces a dipole moment $\vec{d}(t) = \alpha\vec{E}(t)$ in the molecule and the dipole experiences a torque $\vec{\tau}(t) = \vec{d}(t) \times \vec{E}(t)$ [99]. Here, α is the polarizability tensor of the molecule. This torque in turn aligns the molecule, leading to alignment along the polarization axis in the presence of the laser field. The torque is turned on and turned off rapidly during each laser field cycle. For a linear molecule, the polarizability can be described by a polarizability $\alpha_{//}$ parallel to and a polarizability $\alpha_{\perp}$ perpendicular to the molecular axis. The polarizability (β) is defined as the difference between $\alpha_{//}$ and $\alpha_{\perp}$, $\beta = \alpha_{//} - \alpha_{\perp}$. This classical model can be used to explain the alignment of the molecule under the field free condition; however, it is unable to exhibit the revival which has been observed due to the discrete rotational energy levels [103, 104].

In the quantum model, when the molecule experiences an impulse toward alignment, a coherent rotational wave packet is created. Since the laser pulse is much shorter than the rotational period of the molecule, this coherent rotational wave packet is preserved after the laser field is over [105]. The transient alignment periodically occurs due to the dephasing and rephasing of individual components of the rotational wave packet. The degree of alignment strongly depends on the intensity and the pulse duration. There is also a dependence on the laser wavelength; however, it is normally neglected by assuming the field is non-resonant with vibrational or electronic transitions in the molecule [106, 107].

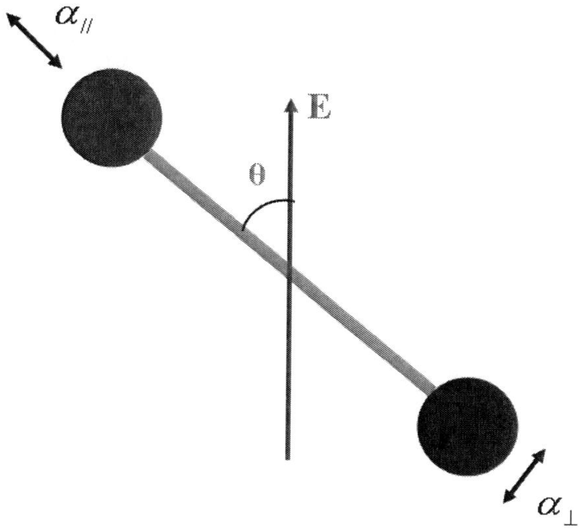

Figure 4.1. The molecule is kicked in the direction of the laser polarization. θ is the angle between the electric field of the laser and the molecular axis.

The rotational wave packet is periodic in time evolution. The period results from rotational eigenstates with discrete energy levels. More details about the rotational wave packet dynamics are presented in the section below.

4.3. Rotational Wave Packet Dynamics

The time evolution of a rotational wave packet originating from the irradiation of a molecule by a laser pulse can be described as the sum over molecular coherence rotational eigenstates [104]

$$\Psi(t) = \sum_{J,M} a_J e^{-i(E_J/\hbar)t} |J,M\rangle \tag{4.1}$$

where $E_J = BJ(J+1)$ are the energy eigenvalues, $B = hcB_0$ is the rotational constant of the molecule which determines the time scale of the rotational dynamics and B_0 is the ground state rotational constant. The $|J,M\rangle$ are the rotational eigenstates, and the M are with respect to the polarization axis of the aligning pulse. The complex coefficients a_J are determined by the interaction potential.

When there is a laser field, the coefficients are time independent. As indicated in the previous section, when interacting with the aligning laser pulse, each constituent rotational state $|J,M\rangle$ accumulates phase with a different angular frequency. These rotational $|J,M\rangle$ states rapidly dephase with respect to each other and the initial net alignment quickly disappears. Since the wave packet evolves in a field-free environment, the $|J,M\rangle$ states eventually start to rephase. The dynamics is due to interference between these states. When the accumulated phases of each $|J,M\rangle$ state are equal, the initial net alignment is reproduced. If one considers the time interval $t = \frac{\pi\hbar}{B}$, then the phase of each state is $\pi J(J+1)$. Since J is a positive integer number, then the phase at this time is a multiple of 2π for any value of J. The wave packet (the alignment) at this time is the same at $t = 0$ (the initial alignment). In other words, the wave packet repeats itself with a period of $T = \frac{\pi\hbar}{B}$ which is called a full revival.

In addition, the selection rules for Raman transitions are $\Delta J = 0, \pm 2$ and $\Delta M = 0$; therefore, at the fractions of this time period, 1/4, 1/2, 3/4, the accumulated phases of states $|J\rangle$ and $|J+2\rangle$ differ by a multiple of 2π [104]. As a result, a significant net alignment and "anti-alignment" are observed at these fractional revivals of molecules with different symmetry.

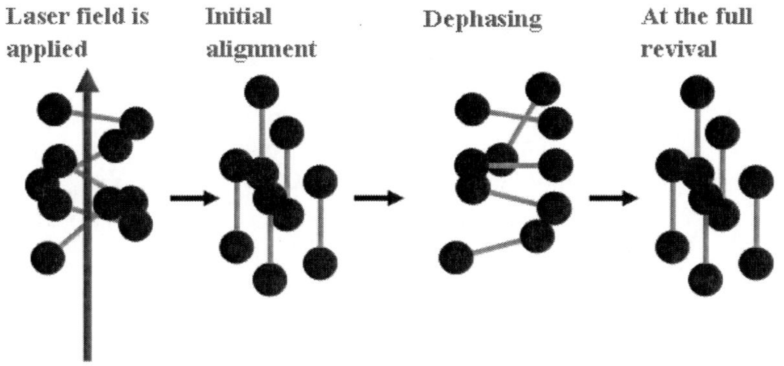

Figure 4.2.Time scale of the non adiabatic alignment process of a molecule.

The degree of alignment is usually characterized by the alignment parameter $\langle \cos^2 \theta \rangle$, where θ is the angle between the molecular axis and the laser polarization direction [99]. $\langle \cos^2 \theta \rangle = 1$ represents ideally complete alignment along the polarization axis, $\langle \cos^2 \theta \rangle = 0$ indicates ideally perfect alignment perpendicular to the polarization axis, and $\langle \cos^2 \theta \rangle = 1/3$ corresponds to an isotropic distribution. Indeed, in the calculation, if $\langle \cos^2 \theta \rangle > 1/3$, the molecule is predominantly aligned along the laser polarization; if $\langle \cos^2 \theta \rangle < 1/3$, the molecule is predominantly aligned perpendicular to the laser polarization [104, 108, 109, 110]. Note that the maximum value and minimum value of $\langle \cos^2 \theta \rangle$ depend on the intensity and the duration of the alignment pulse. In addition to the dependence on the laser parameters, these are also determined by the rotational temperature. Higher temperature creates broader state $|J, M\rangle$ distributions which reduces the net alignment. Therefore, the degree of alignment can be enhanced by reducing the initial temperature of the molecular gas. [99, 111, 112]. The calculated time dependence of the alignment parameters $\langle \cos^2 \theta \rangle$ of nitrogen and oxygen at room temperature are shown below [113], as examples. The particular parameters of these measurements can be found in [113].

The fractional (at T/4, T/2 and 3T/4) and the full revival (T) of the molecular alignment parameter $\langle \cos^2 \theta \rangle$ of N_2 and O_2 are shown in figure 5.3, where T is the natural rotational period of the neutral molecule (for N_2, T ~ 8.4 ps, for O_2, T ~ 11.6 ps) [104]. The initial alignment occurs at t = 0 and is repeated after one full rotational period (8.4 ps for N_2 and 11.6 ps for O_2). There are also half (T/2) and quarter (T/4) revivals for both these diatomic molecules.

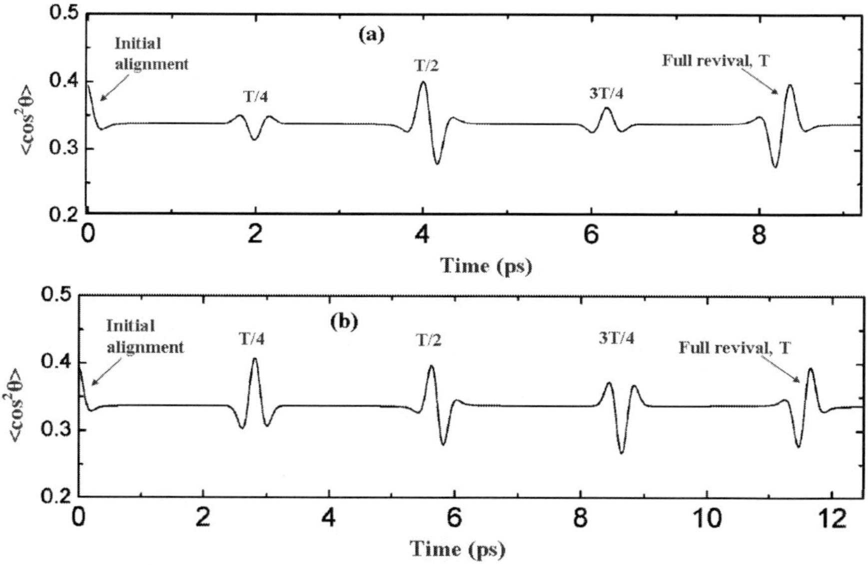

Figure 4.3.Calculated time dependence of the molecular alignment parameter $\langle \cos^2 \theta \rangle$ of nitrogen (a) and oxygen (b).

Comparing figures 4.3a and 4.3b, it is clear that there is different behaviour of the alignment parameter for N_2 and O_2 at the quarter and three-quarter revival. This large difference results from the different weights of the even and odd $|J\rangle$ states for the initial distribution of N_2 and O_2 [104]. At the quarter revival, the phase difference between states $|J\rangle$ and states $|J+2\rangle$ is $\pi/2$. Therefore, at this point, the even $|J\rangle$ states form a secondary wave packet while the odd $|J\rangle$ states form another wave packet. When one secondary packet is predominantly aligned along the laser polarization axis, the other is primarily aligned in the perpendicular direction. In diatomic molecules, the nuclear spin statistics determine the relative weights of the even $|J\rangle$ states and the odd $|J\rangle$ states. For N_2, the relative weights of the even and the odd states are 2:1. Hence, the wave packet formed by the even $|J\rangle$ states is partially cancelled by that formed by the odd $|J\rangle$ states. Consequently, the alignment parameter of N_2 at the quarter revival is smaller than that at the full revival. However, in the case of O_2, in which the nuclear spin statistics allow the population of only odd states, the degree of alignment at the quarter revival is approximately equal to that at the full revival. In addition, due to the different nuclear spin statistics leading to a different ratio of even and odd rotational state populations, the modulation of the alignment parameter of N_2 is out of phase with that of O_2 at the quarter revival.

The base line for $t > 0$ is slightly higher than the isotropic value. This can be explained as follows. For $t < 0$, the relative $|J\rangle$ state populations are given by the thermal Boltzmann distribution and the M values are uniformly populated for each $|J\rangle$ state. The alignment pulse induces rotational excitations and de-excitations via Raman transitions that modify the $|J\rangle$ state distribution but does not change M. In addition, since there is a lower bound for Raman deexcitation, but no such limit for excitation, the wave packet exhibits net rotational excitation. Consequently, the angular momentum vectors for each $|J\rangle$ state in the wave packet are preferentially oriented perpendicular to the alignment pulse polarization. In terms of molecular orientation, this corresponds to a small but persistent net alignment of the ensemble and an elevated experimental baseline which is observed in figure 4.3 [104, 114].

4.4. Molecular Alignment Contribution to Ionization Process

So far, both theoretical [115, 116] and experimental studies [114] have pointed to the fact that the ionization rates depend on the molecular orientation with respect to the laser field. An alignment dependent ionization rate is important for understanding strong field molecular physics. For example, in HHG, the ionization rate plays a crucial role and, therefore, the alignment dependence of the tunnelling ionization probability affects the high harmonic yield. Basically, in diatomic molecules, such as N_2 and O_2, the harmonic intensity varies in phase with the ion yield [92, 95, 117, 118].

For an aligned molecular gas medium, the total ionization signal is given by [114]

$$S \propto \int_0^{2\pi}\int_0^{\pi} P(\theta,\phi)\Pi(\theta)(\sin\theta)d\theta d\phi \tag{4.2}$$

where $P(\theta,\phi)$ is the probability distribution for the molecular orientation with respect to the polarization axis. $\Pi(\theta)$ is the ionization probability for a molecule when the molecular axis is at an angle θ to the laser polarization axis. ϕ is the polar angle about the polarization axis.

Using equation (4.2), I. V. Litvinyuk et al. [114] have calculated the total ionization signal for nitrogen for different angles θ. They found that in N_2 with σ orbital symmetry, the ionization is stronger when aligned along the laser field than when aligned perpendicular to it. For their particular experimental parameters, the ionization probability when $\theta = 0^0$ or 180^0 is approximately four times higher than that when $\theta = 90^0$, which is shown in figure 4.4.

Other clear evidence of alignment dependent ionization probability can be found in [92]. In this research, T. Kansai et al. demonstrated the evolution of the ion intensity from aligned diatomic molecules as a function of the delay time between the pump (aligning) and the probe (driving) beams.

Comparing figures 4.3 and 4.5, it appears that the ion intensity scales linearly with the alignment parameter $<\cos^2\theta>$. This linear dependence can be expressed by [119]

$$\eta \approx \eta_0(0.45 + 1.63 < \cos^2\theta >) \tag{4.3}$$

where η_0 is the ionization fraction for unaligned molecules.

Clearly, for an isotropic distribution $\langle\cos^2\theta\rangle = 1/3$, $\eta \approx \eta_0$.

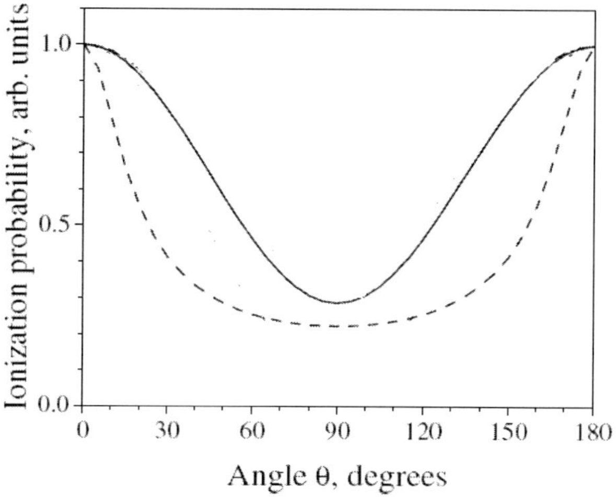

Figure 4.4. Alignment dependence of the ionization probability of nitrogen. The calculation has been done with two possible choices for the monotonic function of θ, $\Pi(\theta)$. $\Pi_1(\theta) = (\cos^2\theta + \varepsilon^2 \sin^2\theta)^{-1/2}$, with $\varepsilon = 4.5$ (long dashed line) and $\Pi_2(\theta) = (\varepsilon\cos^2\theta + 1)/(\varepsilon + 1)$ with $\varepsilon = 3.5$ (solid line). The results are normalized to unity at $\theta = 0^0$.

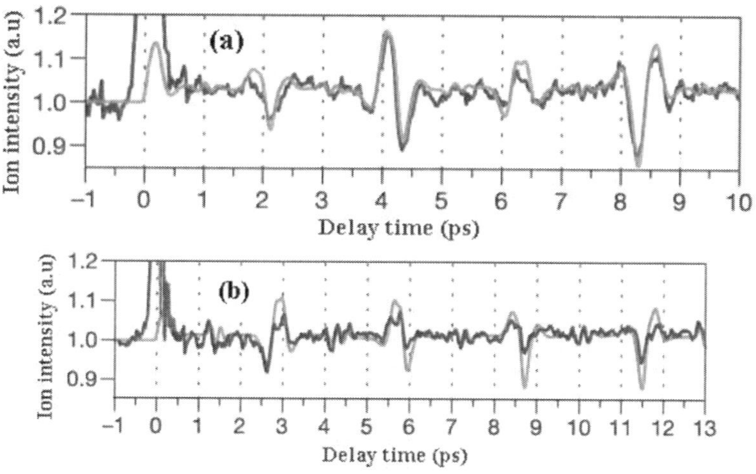

Figure 4.5. Experimental (blue colour) and theoretical (grey colour) results of the ion yield from N_2 (a) and O_2 (b) as a function of time delay between the pump and the probe beams.Positive delay implies that the pump beam precedes the probe beam. The polarization of the pump is set to be parallel to that of the probe beam.

It has been shown that the harmonic intensity in linear diatomic molecules, such as N_2 or O_2,is modulated in phase with the ion rate [92, 95]. By looking at the single-molecule response, one can explain the alignment dependence of the HHG intensity by investigating the angle dependent ionization and recombination processes [92, 95, 109, 117, 118].

The molecular alignment influences the harmonic intensity not only on a single- molecule level, but also on the macroscopic response by changing the properties of the nonlinear medium in which the harmonics propagate, e.g., the refractive index and the free electron density.

4.5. Rotational Raman Contribution to the Nonlinear Refractive Index

When a femtosecond pulse is used to off-resonantly excite linear molecules such as N_2 and O_2,the instantaneous $\Delta n_{inst}(\lambda, t)$ and the non-instantaneous pure rotational Raman $\Delta n_{Raman}(\lambda, t)$ will contribute to the change in total nonlinear refractive index $\Delta n_{total}(\lambda, t)$ [100, 101]

$$\Delta n_{total}(\lambda, t) = \Delta n_{inst}(\lambda, t) + \Delta n_{Raman}(\lambda, t) \tag{4.4}$$

The instantaneous contributions give rise to an instantaneous nonlinearity, in which the magnitude directly follows the intensity profile of the fundamental laser field

$$\Delta n_{inst}(\lambda, t) = n_{2,inst} I(t) \tag{4.5}$$

where $n_{2,inst}$ is the nonlinear refractive-index coefficient due to the bound electrons of the molecules and I(t) is the laser intensity.

The pure non-instantaneous rotational Raman contribution to the change in refractive index is given by

$$\Delta n_{Raman}(\lambda, t) = \mathrm{Im}\left[n_{2,Raman} \sum_{J=0}^{\infty} F_J \times \int_0^{\infty} I(t-\tau)\exp(-i\omega_J \tau) d\tau \right] \tag{4.6}$$

where ω_J is the rotational frequency.

The rotational contribution to the change of refractive index depends on three factors [100, 101]. First, the rotational nonlinear refractive-index coefficient $n_{2,Raman}$ gives the overall magnitude of the contribution independent of the rotational state J of the molecule.

$$n_{2,Raman} = \frac{32\pi^2 N}{hcn_0}\beta^2(\omega) \tag{4.7}$$

where n_0 is the linear refractive index at the laser angular frequency ω, N is the density of the gas, and β(ω) is the anisotrophy of the molecular polarizability as mentioned in section 4.2.1 and given by $\beta = \alpha_{//} - \alpha_{\perp}$

Second, the factor F_J is the weight of the different possible rotational excitations

$$F_J = (\rho_{J+2} - \rho_J)Z_J \frac{(J+2)(J+1)}{(2J+3)} \tag{4.8}$$

F_J is determined by the transition moment between the two coupled rotational levels (J +2) and J and by the population difference (ρ_{J+2} - ρ_J) between the two levels which in turn is determined by the rotational constant of the molecule. The population of a particular level J is

$$\rho_J = \frac{\exp[-hcBJ(J+1)/kT]}{\sum_{J=0}^{\infty} Z_J(2J+1)\exp[-hcBJ(J+1)/kT]} \tag{4.9}$$

where B is the rotational constant of the molecule. Z_J is related to quantum effects due to the nuclear statistics and depends on the molecular symmetry.For N_2, it follows that transitions from even J levels are two times more intense than transitions from odd J levels (Z_J = 1 for odd J and Z_J =2 for even J). For O_2, only odd J levels are populated (Z_J = 1 for odd J and Z_J =0 for even J).

The third factor is the angular frequency difference between the coupled rotational levels (J+2) and J

$$\omega_J = 4\pi cB(2J + 3) \tag{4.10}$$

For gases such as N_2 and O_2, in which the rotational constant B is equal to a few cm^{-1}, the impulsive Raman excitation of rotational coherences can be induced with a single 100 fs pulse. The total contribution of these rotational coherences to the time-dependent change of the refractive index is a sum over all the thermally accessible rotational states and all the optically allowed (Stokes and anti-Stokes) Raman transitions.

Based on the equations above, we can carry out a simulation for the change of the refractive index in N_2 and O_2.

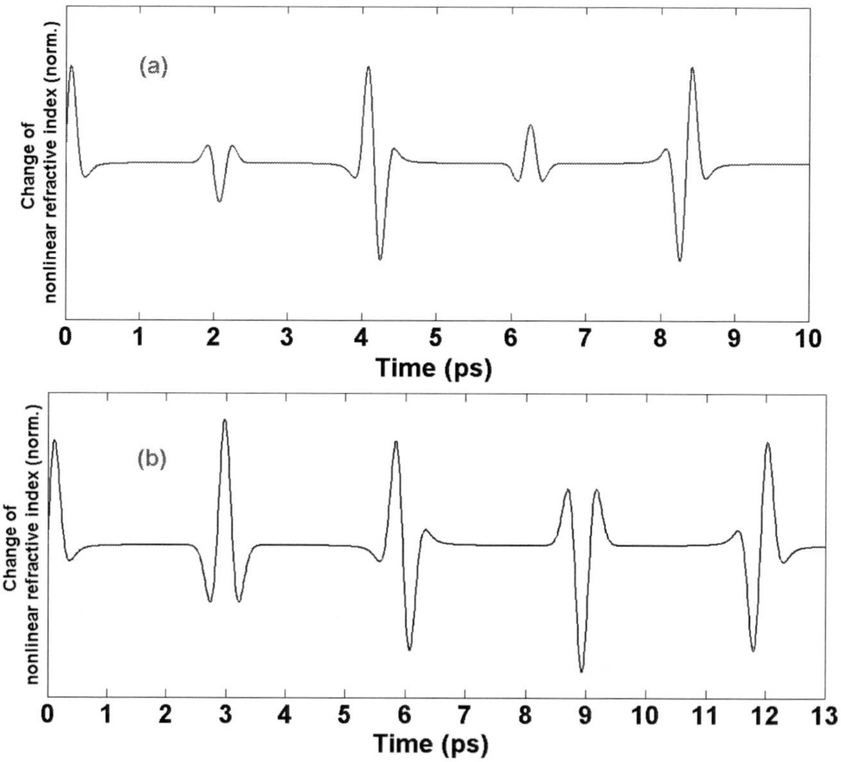

Figure 4.6.Simulation of the normalized change of the nonlinear refractive indexof nitrogen (a) and oxygen (b).

Clearly, comparing figures 4.3 and 4.6, the dephasing and rephrasing of the quantum states induce a period modulation of the nonlinear refractive index of the molecule which is exactly the same as that of the alignment parameter $\langle\cos^2\theta\rangle$. This indicates that the change of the nonlinear refractive index due to the Raman rotational motion Δn_{Raman} is a function of the alignment parameter $\langle\cos^2\theta\rangle$. This relation is shown in [120]

$$\Delta n_{Raman} = \frac{3N_{atm}}{4n_0\varepsilon_0}\beta\left\{<\cos^2\theta> - \frac{1}{3}\right\} \tag{4.11}$$

where N_{atm} is the number density of the gas at 1 atmosphere, ε_0 is the electric constant (vacuum permittivity) and Δn_{Raman} in equation (4.11) is calculated at atmospheric pressure.

In equation (4.11), the term $(<\cos^2\theta> - \frac{1}{3})$ confirms that for the isotropic distribution ($\langle\cos^2\theta\rangle$ =1/3), there is no modulation of the nonlinear refractive index al all.

4.6. Experiment

4.6.1. Experimental Setup

The high power femtosecond laser system and the experimental setup for this study are similar to those described in sections 1.2.1 and 1.2.2. In the two-beam arrangement, the first weak laser beam is used to align the molecules and the second beam generates high order harmonics. The polarization of the first beam relative to that of the second beam can be changed using a half-wave plate. When the first off-axis beam is applied, the molecules are non-adiabatically aligned. To a first approximation the total interaction length can be divided into two regions, one with aligned molecules and the other without.Positive time delay implies that the first (pre-aligning) beam precedes the second (high harmonic generation) beam. Two molecular gases, nitrogen and oxygen, are used in this study.

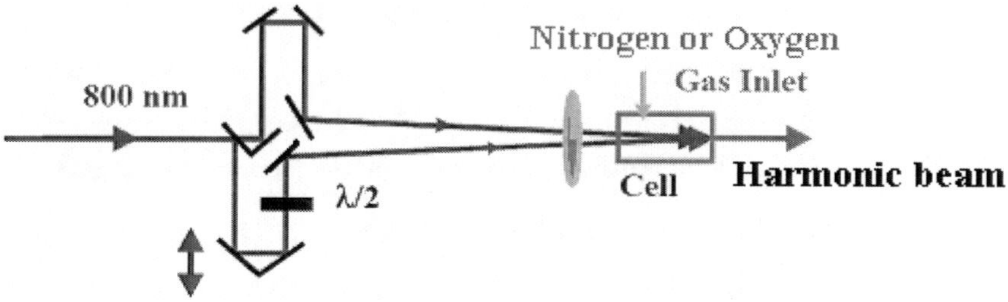

Figure 4.7. Arrangement of the two beams for the rotational molecular dynamics study using a HHG signal.

4.6.2. Results and Interpretation

First, the delay of the first laser beam is set to less than - 2 ps (long negative delay) and for this condition the molecules are not aligned over the total length. The pressure, aperture diameter, and the energy and chirp of the second laser pulse are optimized iterativelyfor maximum flux of all available harmonics. The laser focus is positioned close to the exit of the gas cell. The generation of harmonics is optimized for phase matching along the axis of the second laser beam for the observed spectral range [43].

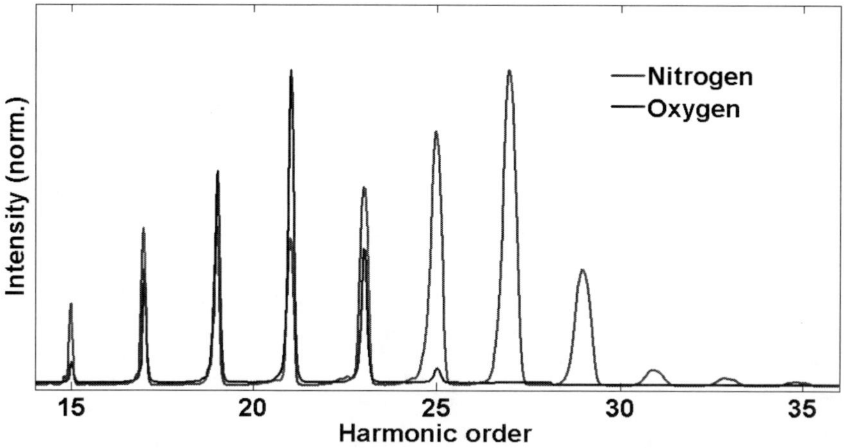

Figure 4.8. Phase Matched optimized HHG from unaligned N_2 (red line) and O_2 (blue line) detected along the axis of the second laser beam.

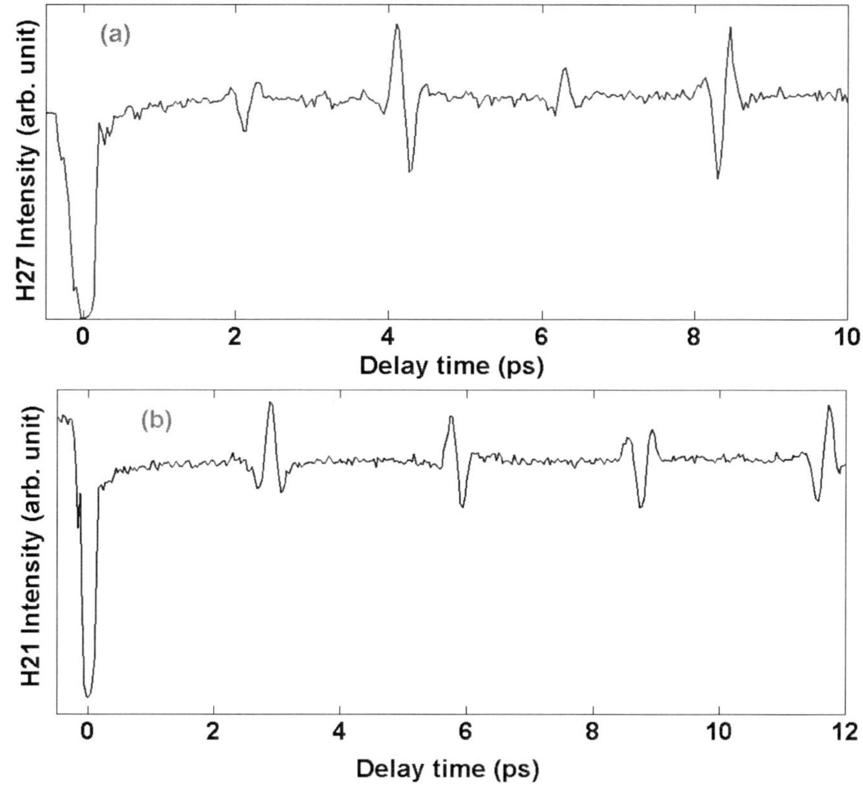

Figure 4.9. Intensity of harmonic H27 for aligned N_2 molecules (a) and H21 for aligned O_2 molecules (b) versus time delay between the two laser beams.

Figure 4.8 shows typical phase matched optimized harmonic generation signals from N_2 and O_2 when there is no pre-aligning pulse and the molecular ensemble is isotropic (unaligned). Clearly, the harmonic orders are higher for N_2 than for O_2 because of the higher ionization energy of N_2.

Figure 4.9 shows the dependence of the HHG intensity on the time delay between the pulses from the two beams for the 27th harmonic order (H27) in N_2 gas (figure 4.9a) and for H21 in O_2 gas (figure 4.9b).

Around zero delay time the harmonic intensities are very low.For a long positive time delay, the harmonic intensity is modulated at delay times around T/4, T/2, 3T/4, where T is the rotational period of the neutral N_2 molecule (T_{N2} = 8.4 ps) or O_2 molecule (T_{O2} = 11.6 ps). Comparing figures 4.9, 4.5 and 4.6, for both N_2 and O_2 it is clear that the modulation of the harmonics in the observed spectral range follows the modulation of the Raman contribution to the nonlinear refractive index and the ion intensity.

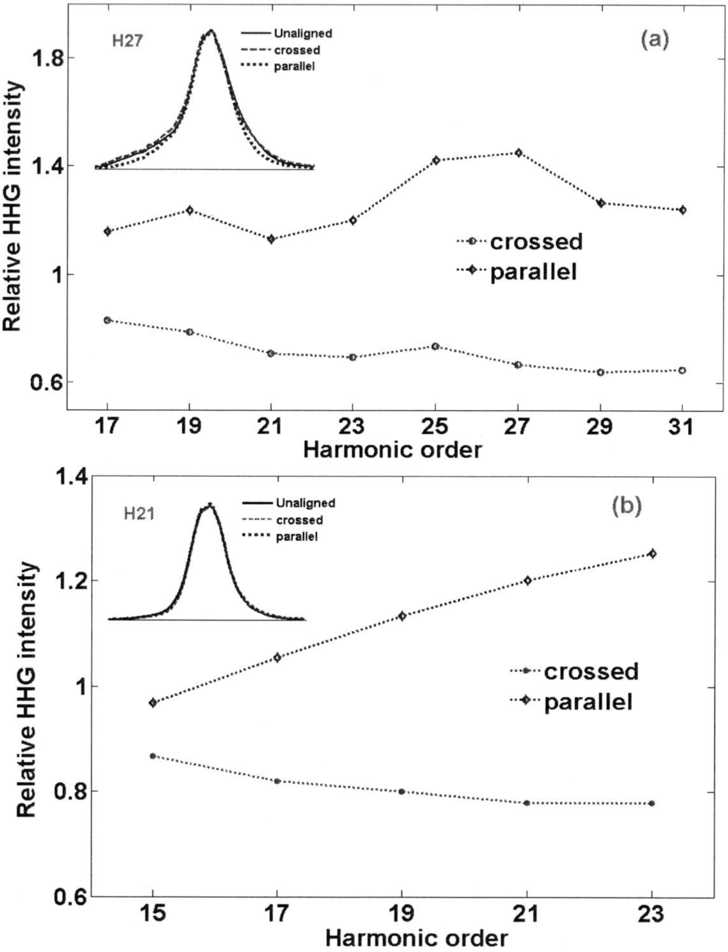

Figure 4.10. Integrated harmonic intensity at delay time T (rotational period) for crossed and parallel alignment normalized to the harmonic intensity of unaligned molecules a) for N_2 and b) for O_2. The insets show the beam profile of (a) the 27th harmonic from N_2 and (b) the 21st harmonic from O_2.

Previous studies by looking at the single-molecule response [96, 97, 117] have demonstrated that the phase and amplitude of the harmonics vary very differently across the HHG spectrum when the molecular alignment is crossed with or parallel to the fundamental field. While the phase of the 17th harmonic from N_2 is not changed for the case of crossed and parallel alignments the phase of the 27th harmonic is changed by ~ $\pi/2$ [96]. This can be understood by investigating the variation of the amplitude and phase of the high harmonic dipole which should be reflected clearly in the profile and spectrum of the harmonics [96]. However, in our study, figure 4.10 shows that the variation of intensity for different harmonics is small and follows a similar trend for crossed and parallel alignment. In addition, the beam profiles are similar for other harmonics. The beam profiles remain unchanged for all harmonics and alignments which indicates that no variation of the dipole phase can be observed during the fractional and full revivals. From this observation, it appears that in our study the macroscopic response through the phase-matching plays the dominant role in the alignment dependence of the observed harmonic intensity instead of the microscopic single-molecule response.

In our experiment, a second beam with high intensity is used to produce HHG over a long interaction medium. Similarly to the experimental setup presented in section 2.2.4, the relatively weak pre-aligning beam used to align the molecules interacts with the driving pulse in only a part of the total interaction length. When the harmonic generation is optimized for phase matching over the total interaction length, a small variation of the total phase matching condition due to the presence of the pre-aligning beam can result in a variation of the harmonic intensity. Indeed, from equation (2.8), which expresses the electric field$E_q(L)$ of the qth harmonic over the length (L) of the interaction medium, and equation (2.11) which represents the total phase mismatch Δk_q over the interaction length L, one finds that if the phase-mismatch is changed step by step in different sections of length L_i the total generated field is given by $E_q(L) = \sum E_q(L_i)$, where the $E_q(L_i)$ can be added to or subtracted from each other depending on their relative phase. This in turn leads to a variation of the HHG signal.

In equation (1.11), since the generating field does not change, the geometric term and the dipole phase term of the total phase mismatch do not change. In addition, as shown in figure 4.10, one can make sure there is no variation of the harmonic dipole in both amplitude and phase. Hence, with the presence of the first beam to align the molecules over only a part of the total interaction length, the total phase mismatch Δk_q can vary because of the change of the first term (medium dispersion) and the second term (plasma dispersion) in equation (1.11). When the molecule is aligned, the variation of the medium dispersion phase mismatch is due to the change of the nonlinear refractive index which has been demonstrated in section 4.4 and the change of the ionization fraction which has been investigated in section 4.5 leads to a variation of the plasma dispersion as well as the medium dispersion phase mismatches. It is worth recalling that the Raman excitation of rotational coherence creates a birefringent medium but only the variation of refractive index along the direction of the electric field of the HHG driving field will influence the propagation of the high harmonic field. Therefore the influence on the harmonic generation for molecules aligned parallel or perpendicular to the driving field is different.

To estimate the degree to which the alignment dependent nonlinear refractive index and ionization fraction influence the total phase mismatch, one can use equations (4.3) and (4.11). N_2 gas, taken as an example of a nonlinear medium, under our experimental conditions similar to those in [121] (pump beam with wavelength 805nm, peak intensity ~ 2×10^{13} W/cm^2,pulse duration ~ 30 fs, at room temperature), we can estimate a maximum value of the alignment parameter $<\cos^2\theta>$ ~0.36. At this value, from equation (4.11), the largest change of the nonlinear refractive index due to the Raman rotational motion at the fundamental wavelength (~805 nm) at atmospheric pressure is approximately 0.63×10^{-5}. Assuming the ionization fraction η is around 4 % when the effective peak intensity at the focus is approximately 10^{14} W/cm^2, from equation (1.11) and with an unchanged geometry term and atomic phase term,one can calculate the variation of the phase mismatch for H27 originating from the change of the nonlinear refractive index $\Delta k(\Delta n)$ ~ 1.5 cm^{-1} (providing that at very high frequency, e.g., the high harmonic frequency, the refractive index is independent of θ [119]). On the other hand, when the alignment parameter reaches the maximum value, from equation (4.3), the ionization fraction increases to 4.2 %. Then, the equivalent variation of the total phase mismatch shown in equation (1.11) $\Delta k(\Delta\eta)$ ~ 2.8 cm^{-1}. It is apparent that these values of $\Delta k(\Delta n)$ and $\Delta k(\Delta\eta)$ are comparable. Therefore, the contribution of both the alignment dependent nonlinear refractive index and the ionization fraction to the change of the total phase mismatch must be considered leading to the modulation of the output harmonic intensity shown in figure 4.9.

When the polarization of the pre-aligning beam is rotated by 90°, the phase of the anisotropy of the molecular polarizability is inverted and therefore the variation of the refractive index along the direction of the HHG driving field is inverted. In addition, the phase modulation of the ion yield is also inverted [92]. Consequently, a modulation inversion of the harmonic intensity is observed when the polarizations are crossed as shown in figure 4.11 for H27 at time T. At the magic angle ($\theta \approx 54.7^{o}$), only an isotropic response is observed. This leads to no change in the refractive index, i.e., there is no modulation of the harmonic intensity.

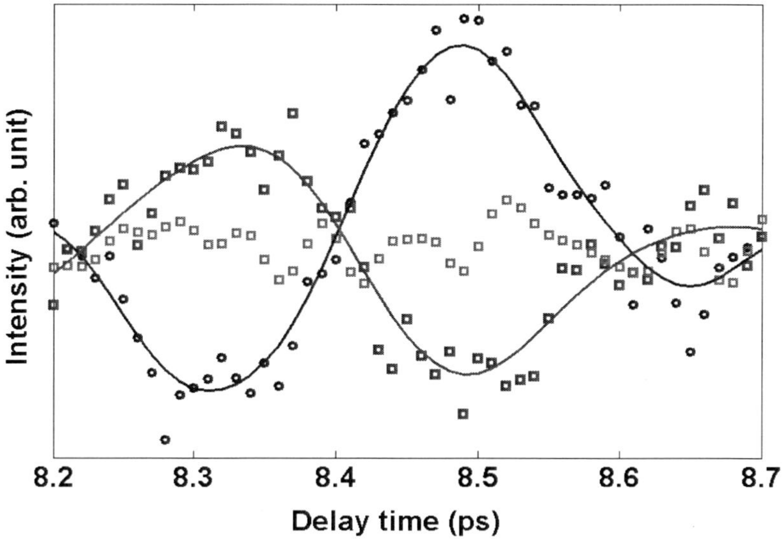

Figure 4.11.Intensity of harmonic H27 for different relative polarizations of the first beam at T (blue – parallel, red – crossed, and green – 54.7^{0}). The symbols are experimental results and the lines represent smoothing curves.

Around zero time delay, the change in the total nonlinear refractive index$\Delta n_{total}(t)$ in the spatial overlapping region is very large because the pre-aligning beam induces both an instantaneous and a non-instantaneous nonlinearity (equation 4.4). The high harmonic dipole phase and the recombination of the electron are strongly influenced by the presence of the first field. The beam profiles and spectrum of the generated harmonics, which reflect the microscopic and macroscopic response, are distorted. The sudden large change of the phase mismatch, Δk_q, in the overlapping region leads to a large phase variation of the integrated harmonic field compared to that in the other region. Therefore, the total phase matching condition is destroyed, i.e., the harmonic intensity is turned off.

Conclusion

In this chapter, we have demonstrated the importance of the effect of phase matching on the HHG signal from aligned molecules. The observed modulations of the high harmonic generation at delay times between the pre-aligning laser beam and the driving beam around T/4, T/2, 3T/4 and T, where T is the rotational period of the neutral N_2 (T_{N2} = 8.4 ps) or O_2 (T_{O2} = 11.6 ps) can be understood by studying the variation of the macroscopic propagation responses which are the medium dispersion and the plasma dispersion. This opens up a new potential technique for studying the coherence dynamics of the ground states of molecules.

Chapter 5

COHERENT DIFFRACTIVE IMAGING USING 13.5 NM HIGH HARMONIC SOURCE

5.1. INTRODUCTION

Microscopy techniques have made many significant achievements with high resolution imaging down to the nanometer scale. Scanning probe microscopy techniques such as atomic force microscopy or scanning force microscopy have been developed with a resolution of the order of fractions of a nanometer [122]. Nevertheless, they are limited to surface structures and are not able to allow us to make a morphological analysis. A well-known breakthrough in imaging technology was the invention of electron microscopy. By using a particle beam of electrons to illuminate a specimen and produce a magnified image, electron microscopy has provided the resolution of crystalline structures at the atomic level [123]. However, this technique is restricted to imaging thin samples. For samples with thickness more than 500 nm, because of inelastic scattering, this technique produces a blurred image and low resolution [124]. To investigate thick samples such as biological specimens, based on the innovation of imaging and labeling techniques, S. M. Hurtley and L. Helmuth developed a far-field light microscope which is able to image living cells with a resolution of about 200 nm [125]. Using extreme ultraviolet or soft x-ray radiation as a light source allows us to obtain high resolution imaging of thick samples [126]. A conventional lens with solid state material which has a strong absorption of soft x-rays cannot be used for XUV radiation focusing and therefore in x-ray microscopy diffractive optical elements such as Fresnel zone plate lenses with very short focal lengths are employed to produce high magnification. By using a condenser zone plate lens to illuminate the sample with x-ray radiation of 1.5 nm wavelength and a micro-zone plate lens to project an image onto a CCD camera, W. Chao et al. [127] achieved a spatial resolution down to 5 nm with a state-of-the-art zone plate. However, for x-ray microscopy, at very high magnification, zone plates have a relatively short working distance and a small depth of focus that leads to the potential for large chromatic aberration. Moreover, since the zone plates block much of the light, there is only around 10% efficiency in the first order diffracted light [126]. Another imaging method with extreme ultraviolet or soft x-ray radiation is holography, such as Fourier transform holography [128]. By using coherent x-ray radiation from a synchrotron, this approach has been used to image gold test objects with a resolution less than 60 nm. However, this method is limited by the need to surround the specimen with a large highly transmitting region.

Recently, coherent x-ray diffractive imaging (CDI) using short wavelength light in the extreme ultraviolet or soft X-ray regions of the spectrum has emerged as a very promising alternative approach for high resolution imaging of thick samples. This technique can overcome some of the shortcomings of the previous methods mentioned above. Especially, CDI is a very useful method to investigate biological samples.

Coherent diffractive imaging is a new imaging technique in which no optical elements such as lenses are needed to obtain the magnified image. CDI was first demonstrated in 1999 using the light from a synchrotron by J. Miao et al. [129]. Instead of using optical elements, in CDI the sample to be investigated is illuminated with a coherent x-ray source. After that, by recording the far-field coherent diffraction pattern, and using iterative phase retrieval algorithms, one can reconstruct the three dimensional image of the object [129, 130, 131]. Since there is no physical lens, this method is free from aberration and provides a very large depth of focus. The resolution of the CDI technique basically depends on the wavelength of the x-rays and the largest angle of scattering where diffraction patterns are recorded. Based on this technique, the structure, shape and size of finite non-periodic objects have been determined [15, 129, 130, 132-136]. Using x-rays from a synchrotron, D. Shapiro et al. have examined a frozen dried yeast cell and its internal components with about 30 nm resolution [137]. Using a soft x-ray free electron laser, H. N. Chapman et al. have successfully imaged a nanostructured non-periodic object with a spatial resolution down to 60 nm [138]. Also, three dimensional reconstruction of objects has been performed successfully at 50 nm resolution [139]. Very recently, C. G. Schroer et al. have shown that by focusing hard x-rays on a sample in the form of a small gold particle, a spatial resolution of about 5 nm can be obtained [136].

Besides the radiation from a synchrotron and from a free electron x-ray laser, High Harmonic Generation (HHG) can be employed as a new x-ray source in CDI. Table top HHG sources for CDI not only provide a high degree of spatial coherence but also enable small scale x-ray microscopy. Due to their nature, HHG sources consist of a beam of multiple longitudinal coherent modes whereas an iterative method for image reconstruction normally needs an entirely coherent wave field. To meet this requirement, i.e., monochromaticity and coherence of the incident x-rays, CDI is usually carried out with a single harmonic order of the harmonic spectrum which is selected by using a monochromator or a narrow band spectral filter. This can lead to a significant decrease of the photon flux of the incident beam on the sample. Alternatively, in past work, to utilize as much of the generated harmonic flux as possible in order to keep the acquisition time short and thus avoid possible long term instabilities of the harmonic source, our collaborators have conducted CDI using multiple order harmonic emission [51, 140, 141, 142]. It has been shown that by employing HHG radiation from an argon source (wavelength ~ 30 nm) with about 5 harmonic orders and a reconstruction using a modified Gerchberg-Saxton iteration algorithm for phase retrieval and a refinement by the application of the maximum entropy method, a spatial resolution of ~ 100 nm has been successfully obtained.

Based on this achievement of CDI using a polychromatic source carried out by our collaborators, the motivation for this work is to pursue a higher resolution of reconstruction. Since the shorter the wavelength of the illumination beam the better the resolution, we will try to perform CDI using a harmonic source based on helium gas with a few harmonics around 13.5 nm focused by a multilayer focusing mirror. A 2D binary pinhole system and a conjugated polymer are used as samples.

5.2. Theoretical Background of Coherent Diffractive Imaging

5.2.1. Introduction to Coherent Diffractive Imaging Technique

When an object of density f(x) is illuminated with a coherent plane wave light field, the diffracted pattern in the far field is the Fourier transform of the function f(x) [143].

$$F(k) = |F(k)| \exp(i\psi(k)) = \hat{F}(f(x)) = \int_{-\infty}^{\infty} t(x) \exp(2\pi ikx) dx \qquad (5.1)$$

Here, x is the three-dimensional spatial coordinate in real space, k is the three- dimensional frequency coordinate in reciprocal space and $\hat{F}$ stands for the Fourier transform operator.

Thus, the Fourier transform F(k) can be written as an amplitude and a phase. In the experiment, only the intensity distribution of F(k), i.e., $|F(k)|^2$, is able to be recorded by the detector and the phase of the pattern is lost. Therefore, in order to fully recover the sample f(x), we must determine the phase information of the function F(k) given by ψ(k).

If both the intensity and phase of the function F(k) are known, the sample can be reconstructed by

$$f(x) = \hat{F}^{-1}\left(\hat{F}(f(x))\right) = \hat{F}^{-1}(|F(k)| \exp(i\psi(k))) \qquad (5.2)$$

where $\hat{F}^{-1}$ denotes the inverse Fourier transform operator.

Briefly, a typical CDI technique includes three basic. Firstly, the sample to be analyzed is illuminated with a monochromatic, coherent wave light field and the diffraction pattern is recorded. Then, the phase of the diffraction pattern is recovered by applying phase retrieval algorithms. Finally, the density of the object in real space is regained from the reciprocal space by using an inverse Fourier transform.

5.2.2. Phase Retrieval

The problem of phase retrieval from the diffraction pattern has been well studied in the field of x-ray crystallography and can be solved by using an oversampling method [144, 145]. To characterize the degree of oversampling, Miao et al. [146] proposed an oversampling ratio, σ, given by

$$\sigma = \frac{(\text{volume of electron density region}) + (\text{volume of no-density region})}{(\text{volume of electron density region})} \tag{5.3}$$

Also, the linear oversampling ratio for two dimensions in reciprocal space, O, is defined as

$$O = \sigma^{1/2} \tag{5.4}$$

Here, if $\sigma > 2$, the number of independent equations exceeds the number of unknowns, the phase information is contained in the diffraction pattern and can in principle be retrieved by algorithms. More details about the algorithms can be found in [147].

Generally, an iterative phase retrieval algorithm consists of the following four basic steps [144, 147] involving iterative Fourier transformation back and forth between the object and Fourier domains and application of the measured data or known constraints in each domain.

1. A random phase set is applied as an initial input. This phase set merging with the magnitude of the Fourier transform, which is calculated by the square root of the measured diffraction intensities, exhibits a function with both amplitude and phase for each pixel.
2. An inverse fast Fourier transform is applied to compute a temporary density function.
3. The new density function recovered in the second step is subjected to the constraints which are positive inside the finite support and zero outside the finite support.
4. A new set of phases can be calculated by employing a fast Fourier transform to the new density function obtained in the third step. The phase of the central pixel is set to zero and this new phase set is used for the next iteration.

The iterations continue until the computed image satisfies the object domain constraints.

6.2.3. Experimental Requirements

In order to achieve a successful and high quality reconstructed image, some experimental requirements need to be met. The sample is illuminated with a plane wave which should be monochromatic and coherent. The spatial coherence length which is defined as

$$L_{S,coh} = \frac{a\lambda}{2\pi d_s} \tag{5.5}$$

needs to be larger than the transverse dimension of the sample. In equation 5.5, a is the distance from the source and d_s is the diameter of the source.

In addition, the coherent photon flux density needs to be high enough to obtain a coherent diffractive pattern at high diffraction angles.

The sample is placed far enough from the detector to ensure that a far-field diffraction pattern is recorded and the Fraunhofer criterion is satisfied.

$$z >> \frac{D^2}{\lambda} \tag{5.6}$$

Here D is the diameter of the sample or the diameter of the illumination area, λ is the wavelength of the illuminating radiation and z is the distance from the sample to the CCD.

From equation (5.4) and the condition $\sigma > 2$ to guarantee that the phase is successfully retrieved from the measured diffraction pattern, the linear oversampling ratio, O, for a two-dimensional object must be [144]

$$O = \sqrt{\sigma} > \sqrt{2} \tag{5.7}$$

In terms of the experimental parameters, the linear oversampling ratio can be expressed as [144]

$$O = \frac{z\lambda}{Dp} \tag{5.8}$$

where p is the pixel size of the CCD camera.

In a successfully reconstructed image, the size of each image pixel is given by [140, 141]

$$r = \alpha\frac{OD}{N} = \alpha\frac{z\lambda}{pN} \tag{5.9}$$

where N is the number of pixels in an N×N CCD array, $\alpha=0.94$ (Sparrow criterion) and $\alpha=1.22$ (Rayleigh criterion).

From equation (5.9), for a given wavelength of the illuminating radiation, the resolution is linearly proportionally to the distance between the sample and the CCD. This means that in order to enhance the resolution of the reconstructed image we can reduce this distance. On the other hand, based on equation (5.7) the ratio between this distance, z, and the diameter of the sample, D, must satisfy

$$\frac{z}{D} > \frac{p}{\lambda}\sqrt{2} \tag{5.10}$$

Therefore, for enhancing the resolution by reducing the distance between the CCD and the sample, one needs to ensure that the diameter of the sample is also reduced so that the conditions for the oversampling ratio and the far field are maintained.

Relating to the requirements of the spectral bandwidth of the illuminating beam for a desired resolution, it has been proposed that [144]

$$\frac{\lambda}{\Delta\lambda} \geq \frac{OD}{d} \tag{5.11}$$

Here, d is the desired resolution of the reconstructed image and $\Delta\lambda$ is the FWHM bandwidth of the incident x-ray.

5.3. Coherent Diffractive Imaging Using 13.5 nm High Harmonic Source

It has been demonstrated that CDI with multiple harmonic orders allows us to keep the acquisition time short and avoid sample damage or detrimental effects due to long term instabilities of the harmonic source [51, 140-142]. This method is well matched to the low temporal coherence and high spatial coherence of the HHG source. A spatial resolution of ~ 100 nm has been achieved by using a HHG source with five harmonics from argon gas for illumination of the sample.

As indicated in equation (5.9), the shorter wavelength of the incident beam can provide a better spatial resolution of reconstruction. Currently, we are able to generate high order harmonics with helium gas in a semi-infinite gas cell. For an optimized observed spectral range we find that the harmonic radiation range around 13.5 nm, which has a short wavelength and high photon flux compared to other ranges, is the most suitable source for the CDI experiments. However, under the best conditions, the number of harmonics from the source is greater than ten, i.e., a ratio $\Delta\lambda/\lambda$ ~ 28 % is observed. Meanwhile, for CDI using a few harmonic source, the maximum value of this ratio must not be higher than 10 % to allow reconstruction of the image [140, 141]. Thus, a multilayer focusing mirror with a narrow bandwidth (FWHM ~ 0.5 nm) around 13.5 nm is installed to provide a narrow spectral range for the illuminating beam. Furthermore, the photon flux generated in helium gas is relatively low compared to that generated in argon gas. Hence, although the harmonic generation around 13.5 nm is optimized leading to the highest achievable photon flux, the intensity of this source is not sufficiently high yet for CDI experiments, i.e., we cannot apply this radiation source to directly illuminate a sample. Therefore, the use of the multilayer mirror focusing to a small point also allows us to significantly improve the photon flux density at the sample.

5.3.1. Experimental Setup and Conditions

In this experiment, the laser system and the high harmonic production and detection systems are similar to those in sections 1.21 and 1.2.2.The essential difference is that in order to acquire the diffraction patterns when the incident x-rays illuminate the sample, the apparatus in the experimental chamber needs to be installed as shown in figure 5.1.

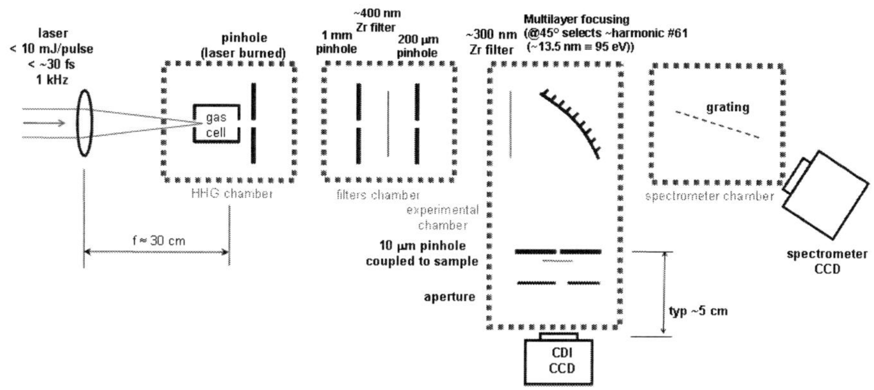

Figure 5.1.Experimental setup for the CDI experiment using a 13.5 nm high harmonic source.

The optimized phase matched HHG radiation is generated from helium gas. To ensure that the fundamental light that co-propagates with the harmonic beam is blocked completely, a second Zr filter (~ 300 nm in thickness) is mounted in the sample chamber. Then, the harmonic beam is spatially filtered by a 200 μm pinhole. The harmonic emission with a few harmonics around 13.5 nm is selected and focused on the sample by a narrow bandwidth (FWHM ~ 0.5 nm) Mo/Si multilayer mirror centred at 13.5 nm with an efficiency of ~ 70%. The installation of the mirror was presented in section 1.2.2.2 and its reflection characteristics are displayed in figure 5.2.

By using this mirror, the beam is focused to a small point which is comparable to the micrometer sized sample, i.e., the size of the illumination source is reduced leading to a sufficiently high photon flux density at the sample with much reduced exposure time.

In addition, this mirror allows narrow bandwidth harmonic emission to be selected. Clearly, according to the characteristics of this mirror shown in figure 5.2, there are a few harmonic orders (~ 3 - 5 harmonics) in the spectrum from 12.5 nm to 14.5 nm to be selected, among which the photon flux of the central harmonic order (~ H59) is strongest and that of the others is much lower. This also leads to the diffraction pattern being an incoherent superposition of the diffraction patterns produced by each harmonic order. However, the contribution of the diffraction caused by the central harmonic dominates; the diffraction caused by the others is minor. The very narrow bandwidth harmonic source selected and focused by the focusing mirror will allow the image of the sample to be reconstructed more easily from the diffraction patterns obtained from the polychromatic source [140, 141].

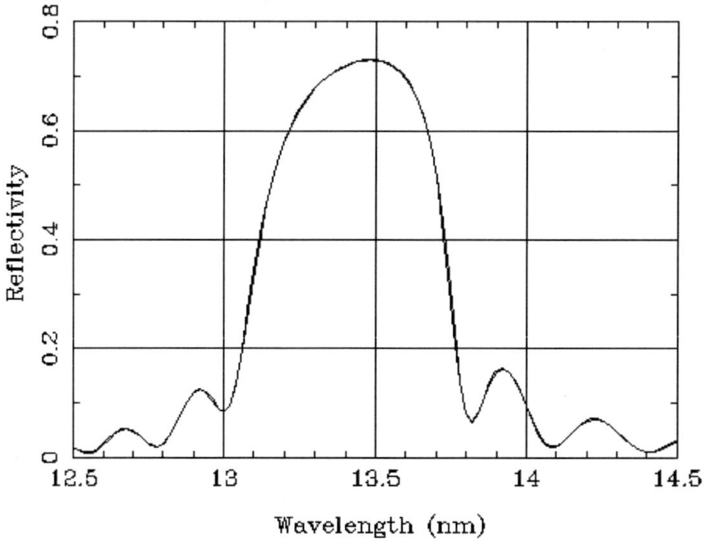

Figure 5.2. Reflectivity of the Mo/Si multilayer focusing mirror (calculated at http://henke.lbl.gov/optical_constants/multi2.html).

The sample is mounted on a holder inside the experimental chamber at a distance of ~ 5 cm far from the focal point of the mirror.Since the focal length is approximately 15 cm, the distance from the mirror to the sample is estimated to be 20 cm. The reason we position the sample at this distance is both to enhance the spatial coherence length and to ensure the photon flux density at the sample is high enough. Equation (5.5) indicates that the further the distance from the sample to the source, the longer is the spatial coherence. However, we need a trade off with the photon flux density at the sample since the further this distance the smaller is the photon flux density illuminating the sample.

In order to guarantee the condition of the oversampling ratio indicated by equation (5.7) to be met and to enhance the coherence of illumination on the sample as well as the contrast of the diffraction pattern at a high angle, a 10 μm pinhole is attached to the sample leading to confinement of the illumination area. In addition, such a determined diameter pinhole resulting in a precise area of illumination can provide a support constraint for the reconstruction algorithm. Obviously, such a pinhole has the potential to introduce distortion to the diffraction image depending on the quality of its edges, and thus to have an influence on the reconstruction. The effects of the pinhole on CDI based on multiple order harmonic emission have been studied by our collaborators in [142].

In addition, when the distance from the focusing mirror and the sample coupled with the 10 μm pinhole is 20 cm, the ratio $\frac{D^2}{\lambda z} = 0.035 << 1$, i.e., the plane wave approximation is satisfied. Thus, the influence of curvature associated with the focusing mirror on the illumination can be ignored and the illumination can be treated as a planar illumination.

The diffraction pattern is captured in the far-field using a CCD camera with 13 μm square pixels on a 1024×1024 array. To satisfy the Fraunhofer criterion of the far- field diffraction the sample needs to be placed at a distance $z >> \frac{D^2}{\lambda} = 7.4\,\text{mm}$ from the detector, where D = 10 μm (confinement of illumination area by using the 10 μm pinhole) is the dimension of the sample. However, as mentioned above, the shorter the distance between the CCD and the sample, the higher the spatial resolution of the reconstructed image that can be achieved. Thus, in practice, the sample is placed at a typical distance of ~ 5 cm from the detector. Consequently, the oversampling ratio is $O = \frac{z\lambda}{Dp} = 5.19 > \sqrt{2}$, so that the phase can be recovered. Based on these experimental parameters, the expected resolution that can be achieved using the Sparrow criterion is givenapproximately by $r = 0.94\frac{OD}{N} = 0.94\frac{z\lambda}{pN} = 50\,\text{nm}$.

To improve the quality of the collected diffraction images by further reducing the scattered light, an aperture with diameter of 5 mm is placed directly downstream from the sample.

To capture a reference spectrum for the reconstruction of the sample, a spectrometer with a grazing incidence diffraction grating and a spectrometer CCD can be inserted into the beam path as shown in figure 5.1.

5.3.2. Samples

In this work, there are two kinds of samples which are investigated. The first sample is a 2D array of regular pinholes and the second is a conjugated polymer film. The installation of the sample holder has been presented in section 1.2.2.2.

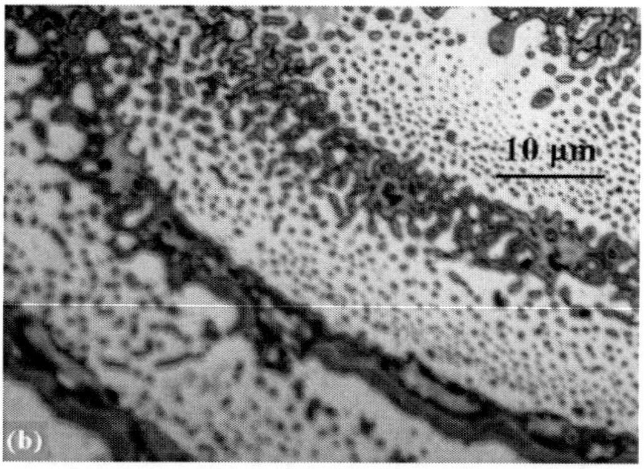

Figure 5.3.Scanning electron microscopy image of the binary regular pinhole sample (a) and optical microscope image of the conjugated polymer sample (b).

The binary regular pinhole sample is micro-machined on a thin (~ 100 nm) gold layer on a Si_3N_4 substrate and consists of an array of 2 μm diameter pinholes spaced by 3 μm and arranged in the domain of a 20 μm diameter circle. This sample does not require a pinhole to restrict the illumination area since it can confine the illumination to a particular area by itself. The sample is either transmissive or opaque to the harmonic emission. It is worth noting that despite not having a pinhole to confine the illumination area the plane wave approximation, i.e., planar illumination, the condition of the oversampling ratio and the Fraunhofer criterion for far-field diffraction are still satisfied when the dimension of the binary pinhole sample is 20 μm and the sample is placed at the same position as stated above (the distance from the focusing mirror and the sample is 20 cm and the distance from the sample to the CCD is 5 cm). A scanning electron microscopy image of this sample is shown in figure 5.3a [142].

The conjugated polymer sample is prepared on a SiN substrate with a thickness of ~ 50 nm. The polymer is MEH-PPV which is purchased from Aldrich and is used without further purification. The powder is dissolved in chlorobenzene with a concentration of 0.1 mg/mL. A total of 100 μL of this solution is then drop-cast onto SiN and the solvent is allowed to evaporate in air for 30 minutes. To confine the illumination area, a 10 μm pinhole is mounted on top of this sample. An optical microscope image of this sample is shown in figure 5.3b.

5.3.3. Generation of the Harmonic Source around 13.5 nm

Before capturing the diffraction images, by properly adjusting the experimental parameters such as the effective interaction length, the chirp of the fundamental laser, the aperture diameter and the gas pressure, the harmonic emission is optimized in an iterative procedure to achieve the phase matched and most intense harmonic orders around H59 (~13.5 nm).For optimized conditions, the pressure of the helium filled gas cell is measured to be ~ 550 Torr, the laser energy is around 2.9 mJ and the focus position is approximately 1 mm inside the gas cell. In our study, use of helium gas as a nonlinear medium in a semi-finite gas cell for harmonic emission in the spectral range between 10 nm and 13 nm provides a higher conversion efficiency than that in the case of neon gas because of its higher ionization energy and since the cross section of helium is low, a high pressure is required.

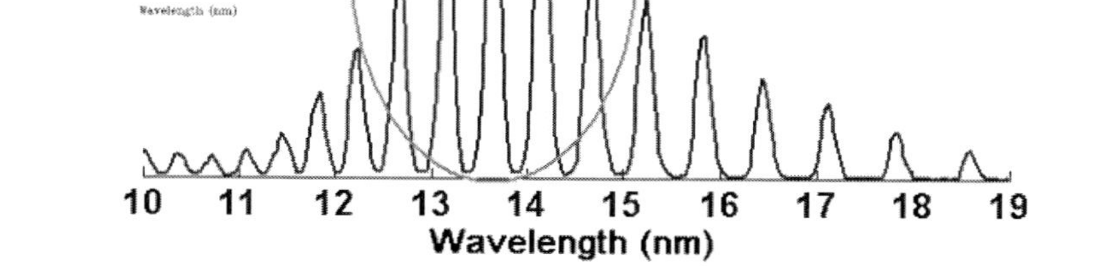

Figure 5.4.Typical optimized harmonic spectrum from helium gas for the CDI experiments (the most intense harmonic emission is around 13.5 nm). The inset is the reflectivity of the focusing mirror as shown in figure 5.2.

Figure 5.4 shows a typical optimized harmonic spectrum from helium gas for our CDI experiments which is recorded by the spectrometer CCD.

When varying the parameters, e.g., the gas pressure, the position of the focus and the aperture diameter, not only the number of harmonic orders but also the spectral weighting within that particular spectral range changes. Thus, for different optimal conditions, the most intense harmonic spectral range can change. For example, if we keep the gas pressure fixed (~ 550 Torr) and vary the aperture diameter and the focus position, the optimized harmonic spectra can exhibit the most intense harmonic orders around 11.5 nm and around 14.5 nm as shown in figures 5.5 and 5.6, respectively. For the spectrum shown in figure 5.5, the focus position is placed at the exit pinhole of the gas cell and the laser energy is increased to ~ 3.2 mJ. For the spectrum shown in figure 5.6, the focus position is moved further into the gas cell at a distance ~ 2 mm from the exit pinhole of the gas cell and the laser energy is reduced to ~ 2.7 mJ.

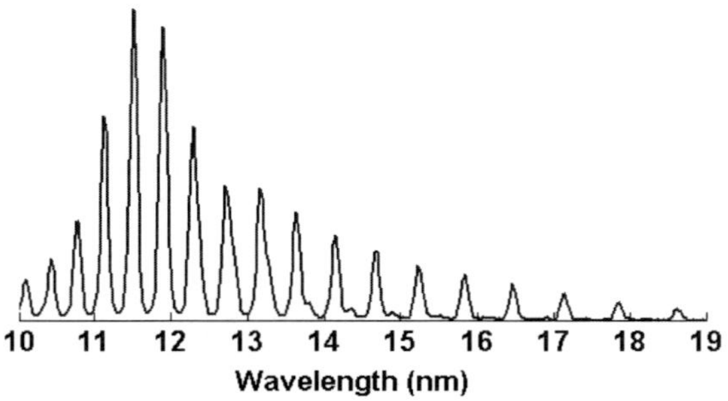

Figure 5.5.Optimized harmonic spectrum from helium gas with the most intense harmonic emission around 11.5 nm.

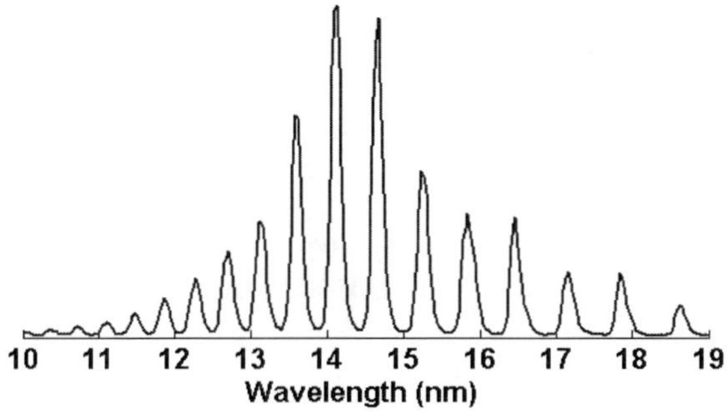

Figure 5.6.Optimized harmonic spectrum from helium gas with the most intense harmonic emission around 14.5 nm.

From figures 5.4, 5.5 and 5.6, it is clear that by varying the experimental parameters such as the gas pressure, the focus position and the aperture diameter, we are able to obtain harmonic radiation with different maximum intensity ranges over the harmonic spectrum. However, for this study, only the harmonic spectrum with the most intense emission around 13.5 nm (shown in figure 5.4) is the most suitable.

5.3.4. Spatial Coherence of the Harmonic Source

As explained in section 5.2.3, one of the critical requirements of the CDI technique is spatial coherence. It is expected that the spatial coherence length which is defined by equation (5.5) needs to be larger than the transverse dimension of the sample. Moreover, for the multiple harmonic order CDI reconstruction algorithm modified by the Gerchberg-Saxton iteration, a high spatial coherent illumination source is also required. Thus, in the following, the degree of spatial coherence of the harmonic source with a few harmonics around 13.5 nm is determined and discussed.

5.3.4.1 Degree of Spatial Coherence of the Harmonic Source

In order to determine the degree of spatial coherence, a Young's double slit which has slit spacing d, slit width w, and slit height h, is used. Choice of an appropriate value of h can ensure that the interferogram covers a reasonable number of pixels on the CCD in a direction perpendicular to the interference pattern [43, 44].

For a typical measurement of the coherence property of a monochromatic electric field, by illuminating a double slit, the fringe visibility is given by

$$\vartheta = \frac{I_{max} - I_{min}}{I_{max} + I_{min}} \tag{5.12}$$

where I_{max} is the intensity of the centre and I_{min} is the intensity of the first minimum of the interference pattern. The source would be perfectly spatially coherent when $\vartheta = 1$.

The degree of spatial coherence is related to the fringe visibility and is expressed as

$$|\gamma_{12}| = \vartheta \frac{I_1 + I_2}{2\sqrt{I_1 I_2}} \tag{5.13}$$

where I_1 and I_2 are the intensities of the electric field at the two individual slits of the Young double slit. Thus, the degree of spatial coherence not only depends on the visibility but also on the strength of illumination on both individual slits. To simplify the measurement, in the case when the slits are illuminated equally, i.e., $I_1 = I_2$, the degree of spatial coherence is given directly by the visibility $|\gamma_{12}| \approx \vartheta = \frac{I_{max} - I_{min}}{I_{max} + I_{min}}$.

In the case of monochromatic illumination with wavelength λ_q and a degree of spatial coherence$\gamma_{12(q)}$, the measured intensity distribution can be written as

$$I(x \mid \lambda_q) = 2\left[1 + |\gamma_{12(q)}| \cos\left(2\pi \frac{xd}{cz\lambda_q}\right) \text{sinc}^2\left(\pi \frac{w}{z\lambda_q}\right)\right] \tag{5.14}$$

where x is the spatial coordinate in the detector plane, c is the speed of the light, d and w are the slit spacing and the slit width, respectively, and z is the distance from the Young's double slit to the detector plane.

For a polychromatic source centred at wavelength λ_q, the measured intensity distribution of the interference pattern on the CCD is given by

$$I_T(x) = \sum_q F(\lambda_q) I(x \mid \lambda_q) \tag{5.15}$$

where $F(\lambda_q)$ is the relative weighting of the qth harmonic order in the power spectrum.

In this case, the average degree of spatial coherence can be determined by

$$\langle |\gamma_{12}| \rangle \approx \langle \vartheta \rangle = \frac{I_{T,max} - I_{T,min}}{I_{T,max} + I_{T,min}} \tag{5.16}$$

In the case of a few generated harmonic orders, we can assume that the harmonic orders have the same or similar degree of the spatial coherence. In fact, the measured average degree of spatial coherence varies little when changing the experimental parameters which strongly influence the number of emitted harmonic orders and their relative weighting in the spectrum [43, 44]. Consequently, the degree of spatial coherence is determined by an average value of all available harmonic orders in the spectrum and is calculated from the fringe visibility of the measured interference pattern.

5.3.4.2. Measurement of Degree of Spatial Coherence

In our study, we use a Young's double slit with a slit spacing of 15 μm, a slit width of 1 μm and a slit height of 50 μm to determine the degree of spatial coherence of the harmonic source around 13.5 nm, i.e., the harmonic radiation after the focusing mirror. The Young's double slit is etched into a silicon wafer which is mounted on the slit holder.

The laser system is described in section 1.2.1. The experimental setup is similar to figure 5.1 in which the slit holder is glued onto the sample holder instead of the sample. The sample holder is then placed on a linear stepper motor which is mounted on a set of x/y-translation stages for precise control of the three directions, x, y and z.

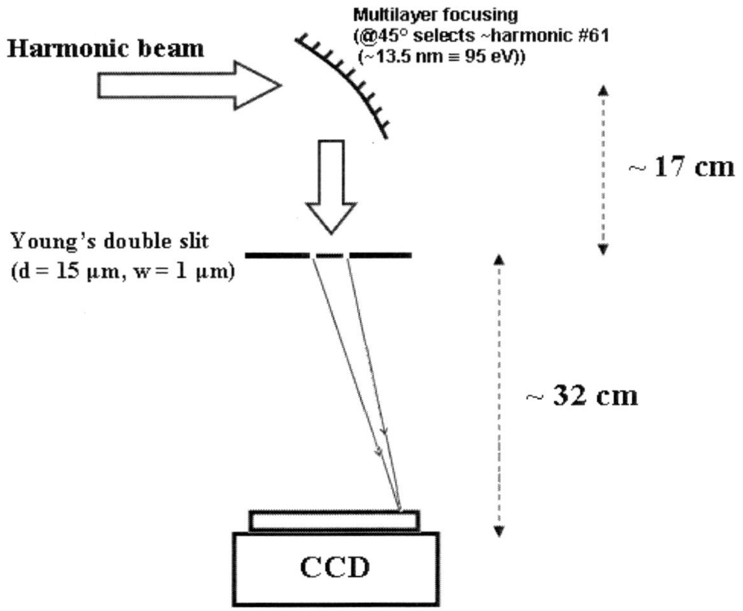

Figure 5.7.Scheme of the Young's double slit experiment.

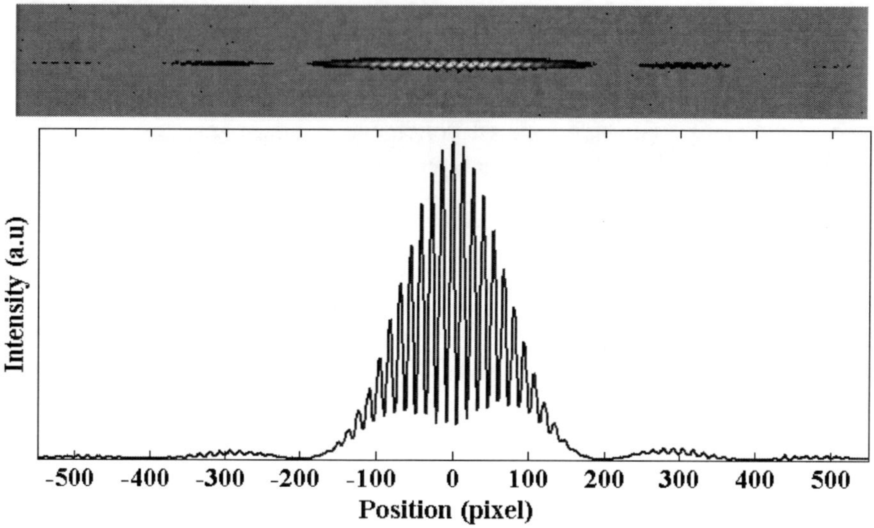

Figure 5.8. Interferogram recorded in the CCD (top) and the interference pattern (bottom) when the 15 μm separation Young's double slit is used.

In fact, we would also like to measure the degree of spatial coherence on different length scales by using different Young's double slits. Unfortunately, the other double slits were destroyed during the fabrication process and thus only a double slit with a spacing of 15 μm was available for our experiment. The distance between the Young's double slit and the focusing mirror can be controlled by moving the linear stepper motor. In order to obtain reasonable photon flux illumination, the slit is placed close to the focusing point of the mirror (2 cm from the focusing point), i.e., the distance from the mirror and the slit is around 17 cm (the focal length of the mirror is about 15 cm). In addition, to increase the number of experimental points in the interference pattern for the reconstructed spectrum (presented in the next section) the distance between the Young's double slit and the CCD needs to be lengthened. In our experiment, this distance is approximately 32 cm.

Figure 5.8 shows the interferogram recorded by the CCD and the interference pattern when the Young's double slit with 15 μm separation is used.

Based on the interference pattern shown in figure 5.8, the degree of spatial coherence of this source is ~ 0.81. This value is reasonably high and indicates that the harmonic source around 13.5 nm generated from a helium semi-infinite gas cell can be used for multiple wavelength coherent diffractive imaging.

5.3.5. Spectrum Reconstruction from Young's Double Slit Interference Pattern

In order to recover the object density from diffraction patterns created by a polychromatic source, a reference spectrum must be known, i.e., the spectral power distribution $F(\lambda_q)$ shown in equation (5.15) is characterized. In the reconstruction algorithm, extraction of the spectral power distribution harmonic beam is used as a starting point to fit the predicted intensity distribution and the measured intensity distribution via the distribution of spectral weights [140, 141]. In previous works, our colleagues have also successfully recovered the HHG spectrum from argon gas from a fringe pattern created by a Young's double slit by application of the maximum entropy method (MEM) [148, 149]. A detailed description of the maximum entropy method is presented in [150]. The general idea of a MEM based reconstruction is that the most probable HHG power spectrum distribution is given by the distribution with the maximum value of the entropy functional [148]

$$S(\lambda)^{n+1} = S(\lambda)^n \exp\left(-\eta \frac{\partial C}{\partial S}\right) \tag{5.17}$$

which satisfies the following condition

$$C = \frac{1}{M}\sum_{j=1}^{M}\left(\frac{\left|I_T^{MEM}(x_j) - I_T(x_j)\right|^2}{\sigma^2(I_T)}\right) \tag{5.18}$$

where η is the Lagrange multiplier, M is the number of the experimental points, $I_T(x_j)$ is the experimental intensity distribution of the interference pattern, $I_T^{MEM}(x_j)$ is the maximum entropy method estimate of $I_T(x_j)$ and $\sigma(I_T)$ is the estimated uncertainty in the measurement of I_T.

In order to evaluate the reconstruction process of the spectral characteristics, the reliability is given by

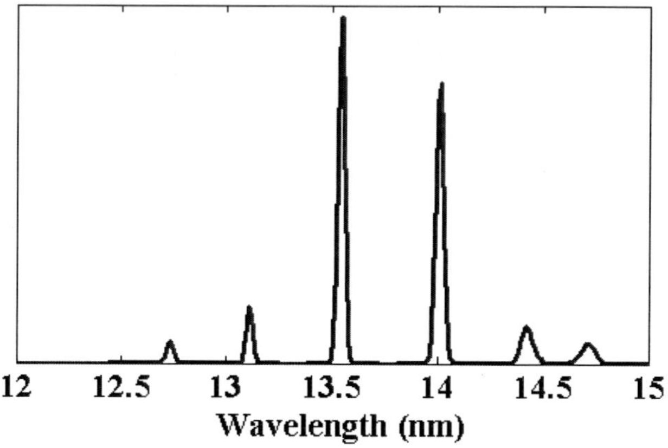

Figure 5.9.Reconstructed spectrum of the harmonicsource around 13.5 nm.

$$R_F(\%) = 100\left(1 - \frac{\sum_{j=1}^{M}\left|I_T^{MEM}(x_j) - I_T(x_j)\right|}{\sum_{j=1}^{M} I_T(x_j)}\right) \tag{5.19}$$

When $R_F > 95\%$, the MEM based spectrum reconstruction is considered to be successful and reliable [148].

In this study, we also apply the MEM to obtain the reconstructed harmonic spectrum from the interferogram shown in figure 5.8.

Clearly, the reconstructed spectrum with the most intense harmonics around 13.5 nm is, in general, consistent with that which would be observed when the full range of the optimized spectrum is reflected from the Mo/Si multilayer focusing mirror with the characteristics shown in figure 5.2. However, the reliability of the reconstruction process is ~ 90 % which is still insufficient for the reconstructed spectrum process. This problem may be due to the inaccuracy of the experimental parameters such as the width of the Young's double slit, the angle of the interference pattern with respect to the CCD pixel array, the slit separation and the distance between the CCD and the Young's double slit all of which strongly influence the reconstructed spectrum. It has been shown that even a very small variation (< 1 %) in these parameters can lead to remarkably inaccurate relative weighting and shifting of the reconstructed harmonic orders [51].

A more important problem is that the required distance between the CCD and the sample in the CDI experiments (~ 5 cm) is very different from that between the CCD and the slits in the Young's double slit experiments (~ 32 cm). Thus, we are not able to acquire the diffractive images and the interference image for spectral characterization of the incident harmonic beam under the same experimental condition in order that the spectrum extraction could be used as a reference spectrum in the reconstruction algorithm. To deal with this problem, a flexible extension tube which connects the experimental chamber to the CCD and allows us to vary the distance between the CCD to the sample (or the slit) easily could be installed.

In this study, to obtain the reference spectrum for the reconstruction algorithm without the use of a Young's double slit, a spectrometer chamber and a spectrometer CCD are implemented as shown in figure 5.1. When the focusing mirror is lifted, the full range of the harmonic spectrum, which is shown in figure 5.4, can be measured. After that, based on the characteristics of the mirror shown in figure 5.2, a reference spectrum of the harmonics around 13.5 nm could be estimated for the reconstruction algorithm.

5.3.6. Acquisition of Diffractive Images

Firstly, macroscopic parameters such as the gas pressure, the energy and chirp of the laser, and the position of the laser focus relative to the exit of the gas cell are optimized in order that the phase matching and maximum photon flux of the harmonic orders around H59 (~13.5 nm) are achieved.

The multilayer focusing mirror is moved into the beam path so that the harmonic emission from around 12.5 nm to 14.5 nm, in which the photon flux of the central harmonic order ~H59 is strongest, is focused onto the sample. Then, the sample is placed so that the centre of the diffraction pattern coincides with the centre of the detector (CCD) which is cooled to -30^{0}C to equilibrate thermally.

In order to achieve a high resolution of the reconstructed sample from the coherent diffractive image, diffraction images with high angle diffractive features which can provide high resolution information need to be captured with a long exposure time. However, due to the limitation of the dynamical range of the CCD (~ 60,000 counts), the diffractive features in the centre or at low angles saturate rapidly when a long exposure time is required. Therefore, we can perform this task by taking data frames at different exposure times. First, a data frame is taken with a appropriately short exposure time which ensures that all pixels in the CCD are not saturated. This short exposure time does not allow the acquisition of high angle diffraction data; only the low scattering angle data, particularly the features at the centre of the diffractive image, have been collected. Then, the exposure time is increased step by step so that the higher angle diffraction data is collected. This acquisition process is stopped when the highest possible angle diffraction pattern is obtained. With a longer exposure time, the low scattering angle data is highly overexposed and the CCD is saturated in these regions. Eventually, these data sets are normalized to each other and then combined so as to obtain the high angle scatter from the long exposure data with the non-saturated data together from the shorter exposure time data. It is advised that during the acquisition process the exposure time should not jump sharply since this may cause a sudden broad saturation region; therefore, some diffractive features would be lost when the images are stitched together. In order to eliminate any ghost image due to remaining charge from the previous image on the CCD detector chip, the detector system should be powered down and restarted after each signal image is taken. We also note that each of the signal images consists of several frames for averaging and need to have their background subtracted. The background images are dominated by thermal and readout noise and are taken by venting the gas cell.

5.3.7. Imaging Results

The diffraction image of the 2D regular pinhole sample is obtained by combining the acquisition of five different images taken at different appropriate exposure times. For the shortest exposure time, 3s, the image features a region of interest (ROI) of low angle diffraction data in its centre as shown in figure 5.10a. This exposure time ensures that there is no pixel in the CCD reaching saturation. After that, the exposure time increases gradually from 3 seconds to 30 seconds (figure 5.10b), 3.5 minutes (figure 5.10c), 7 minutes (figure 5.10d) and 12.5 minutes (figure 5.10e) for acquisition of the high angle diffraction information. As mentioned above, if the exposure time changes abruptly, for example, from 3 seconds to 7 minutes, and then the two signal images (shown in figures 5.10a and 5.10d respectively) were combined to remove saturation regions, we may lose the diffractive features in the marked rectangular areas in figure 5.10d.

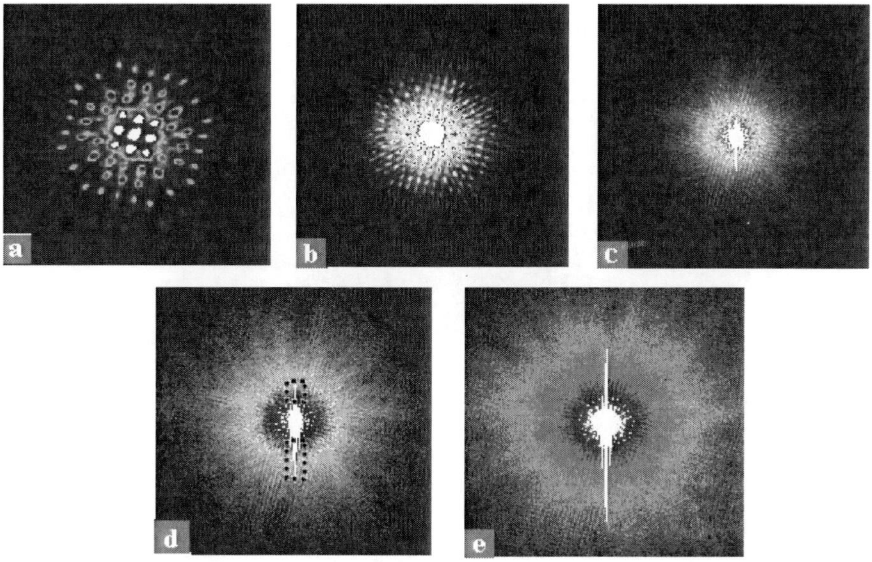

Figure 5.10. Diffraction images from the 2D regular binary pinhole sample at different exposure times: 3 seconds (a), 30 seconds (b), 3.5 minutes (c), 7 minutes (d) and 12.5 minutes (e). The dashed rectangles in figure 6.10d denote areas where the diffraction features may be lost if images (a) and (d) were to be combined directly.

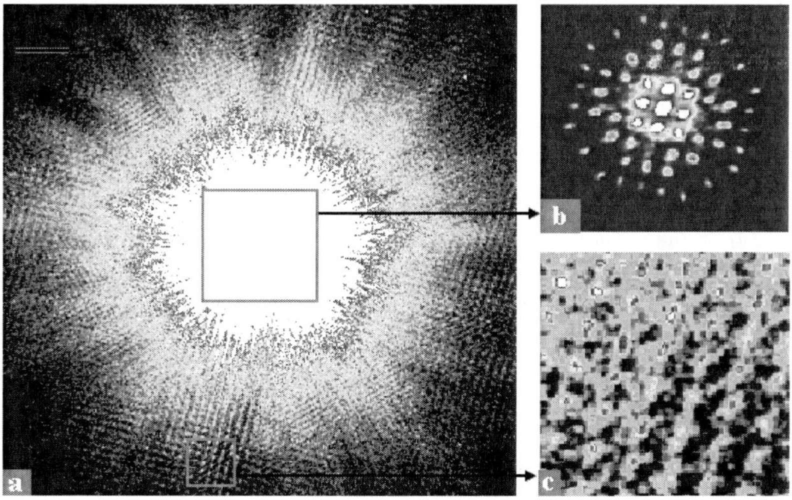

Figure 5.11. Diffraction image of the 2D regular pinhole sample (a), zoom of the centre area (b) and zoom of a high-angle diffraction area (c).

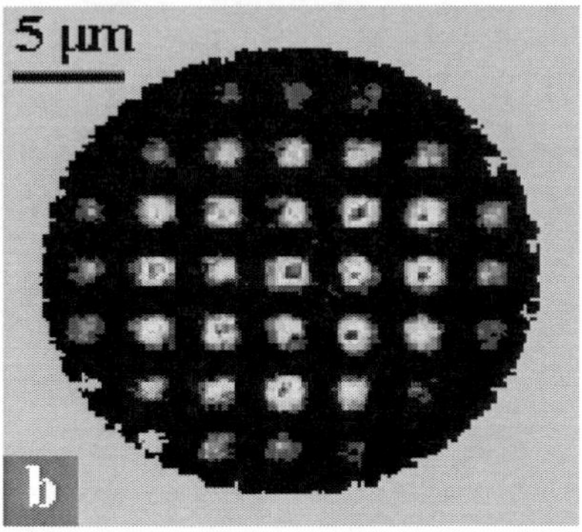

Figure 5.12. Fourier transform of the diffraction image (a), and the low resolution and high noise reconstructed image of the 2D regular pinhole sample (b).

By combining the five different diffraction images shown in figure 5.10, the final diffraction image of the 2D regular binary pinhole sample which comprises both low- and high-angle diffraction data without any saturation is achieved and presented in figure 5.11a. Figures 5.11b and 5.11c show a zoom of the centre diffraction area and the high angle diffraction area. The diffraction image has a strong qualitative resemblance to the pattern that would be produced by an extended periodic sample. The Fourier transform of the diffraction image is shown in figure 5.12a and the low resolution, high noise reconstructed image of the 2D regular pinhole sample is presented in figure 5.12b. The modified Gerchberg-Saxton iteration algorithm, which was developed by our colleagues for the multi-wavelength CDI approach, is used to obtain this reconstructed image. A detailed description of this algorithm can be found in [140, 141]. This algorithm does not require an illumination wave field that exhibits both perfect spatial and temporal coherence and allows diffraction images of a regular object to be processed that is a superposition of different diffraction images produced by each harmonic order of the polychromatic source. In this iterative algorithm, the exit wave field $G(\rho_0 \mid \lambda_k)$ leaving an object can be written in terms of a thickness function, $T(\rho_0)$, where ρ_0 is the position vector in the object plane and λ_k is the wavelength of the kth mode after we sample the measured spectrum into a set of N values S_k, $k = 1, 2, \ldots, N$. The initial guess for the object density is obtained from the autocorrelation function of the image obtained by Fourier transforming the measured diffraction intensity. The key idea of CDI using a polychromatic source is to reconstruct the wavelength independent thickness function $T(\rho_0)$. Each of the modes must individually satisfy the support constraint but only the most central mode in the spectrum $G(\rho_0 \mid \lambda_c)$ is propagated back to the sample and that is used as a basic for the update of the estimate $T(\rho_0)$ in the next iterative cycle. Thus, the key extension of this algorithm compared to the conventional Gerchberg-Saxton iteration algorithm is a combination of multiple forward modes and a single back-propagation mode. It is useful to recall that a high spatial coherence is required for accuracy with this algorithm.

Figure 5.13a shows the diffraction image of the conjugated polymer sample and figures 5.13b and 5.13c are zooms of the centre area and the high-angle area. Similar to the case of the 2D regular pinhole sample, the diffraction image of the conjugated polymer sample also originates from a combination of six sets with different exposure times: 5 seconds, 15 seconds, 45 seconds, 2 minutes, 5 minutes and 10 minutes. Since this is a random and complex sample, its diffraction pattern is much more complicated than that from the 2D regular pinhole sample. Therefore, for successful recovery of the density of the complex sample, a new reconstruction algorithm which allows the diffraction patterns of the random (non-homogeneous) sample to be processed needs to be developed by the theoretical group.

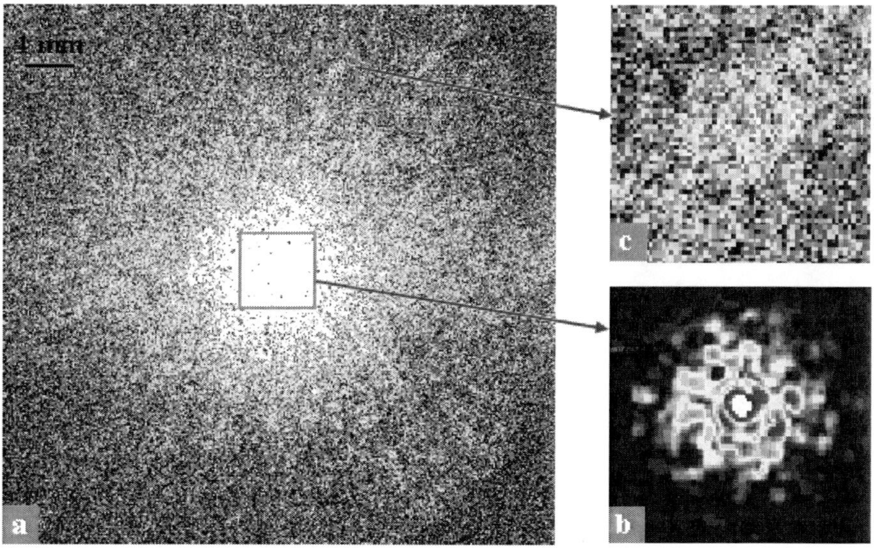

Figure 5.13. Diffraction image of the conjugated polymer sample (a),zoom of the centre area (b) and zoom of a high-angle diffraction area(c).

5.3.8. Limitations

So far the high resolution of the reconstructed image of the binary pinhole sample could not be achieved because of some limitations. The first limitation originates from the spatial coherence of the illumination source. As analyzed in section 5.3.4, the degree of spatial coherence of the harmonic beam consisting of a few harmonics around 13.5 nm is ~ 0.81 which is rather lower than that of the harmonic source generated from argon (~ 0.95) [44, 149]. This may result from the fact that the spatial coherence length is shorter than the transverse dimension of the sample used in this study (20 μm) leading to some distortion in the diffraction images. The second limitation is the instability of the incident harmonic beam during the long exposure time required to overcome the low photon flux. This may be due to the instability of the fundamental laser beam. Since the harmonic signal is generated from helium gas with a very high ionization potential, even a small variation of the beam profile of the fundamental laser can cause a significantly unstable harmonic beam. This problem is responsible for the blurring of the diffraction images which can be seen clearly in figure 5.10a. Also, regarding the experimental setup described in figure 5.1, we know that when the mirror is inserted at an incident angle of 45^0 it could not provide a good focusing of the harmonic beam. Thus, a long exposure time for acquisition of high-angle diffraction data is still required. At this point in time, a reconstruction algorithm for a random sample is still not available; therefore, no reconstructed image of the conjugated polymer is shown.

In order to obtain a high quality reconstructed image with high spatial resolution, we need to solve these problems. Firstly, the stability of the source could be significantly enhanced by employing an auto-stabilization system, which is currently under construction in our laboratory, to keep the fundamental beam profile unchanged. Secondly, the spatial coherence of the harmonic generation around 13.5 nm generated from helium gas needs to be improved so that the coherence length is longer than the sample size. Furthermore, the experimental setup needs to be rearranged so that the incident angle is smaller; in this way, the focusing of the harmonic photons could be stronger in order to improve the photon flux illuminating the sample for reduction of the exposure time.

CONCLUSION

We have demonstrated the possibility of using a x-ray source around 13.5 nm based on the harmonic emission from helium gas for coherent diffractive imaging to improve the spatial resolution of reconstructed images. Harmonic emission with the most intense harmonic orders around H59 (~ 13.5 nm) has been observed and by using a Young's double slit the degree of spatial coherence of this source was measured to be ~ 0.81 which is reasonably high for multiple wavelength coherent diffractive imaging. The diffraction patterns of a regular object as well as a complex object have been recored. Furthermore, a low resolution and high noise reconstruction of the 2D binary pinhole sample has been obtained. However, so far, we have not been able to recover the object density with high spatial resolution. This is due to the poor qualities of the harmonic source involving the spatial coherence and the stability. In order to be successful with a high resolution output of the reconstruction process, apart from enhancement of the qualities of the incident harmonic beam and rearrangement of the experimental setup for efficient focusing of the mirror, new reconstruction and refinement algorithms for the complex and random samples need to be developed by the theoretical group.

References

[1] P. A. Franken, A. E. Hill, C. W. Peters, and G. Weinreich, "Generation ofoptical harmonics," *Physical Review Letters* 7, 118 (1961).

[2] G. H. C. New, and J. F. Ward, "Optical third-harmonic generation in gases," *Physical Review Letters* 19, 556 (1967).

[3] J. L. Krause, K. J. Schafer, and K. C. Kulander, "High-order harmonic generationfrom atoms and ions in the high intensity regime," *Physical Review Letters* 68, 3535 (1992).

[4] J. J. Macklin, J. D. Kmetec, and C. L. Gordon, III, "High-order harmonic generation using intense femtosecond pulses," *Physical Review Letters* 70, 766 (1993).

[5] A. McPherson, G. Gibson, H. Jara, U. Johann, T. S. Luk, I. A. McIntyre, K. Boyer, and C. K. Rhodes, "Studies of multiphoton production of vacuum-ultraviolet radiation in the rare gases," *Journal of the OpticalSociety of America* B 4, 595 (1987).

[6] A. L'Huillier, K. J. Schafer, and K. C. Kulander, "Higher-order harmonic generation in xenon at 1064 nm: The role of phase matching," *Physical Review Letters* 66, 2200 (1991).

[7] A. L'Huillier, P. Balcou, S. Candel, K. J. Schafer, and K. C. Kulander, "Calculations of high-order harmonic-generation processes in xenon at 1064 nm," *Physical Review A* 46, 2778 (1992).

[8] P. B. Corkum, "Plasma perspective on strong field multiphoton ionization," *Physical Review Letters* 71, 1994 (1993).

[9] J. Zhou, J. Peatross, M. M. Murnane, H. C. Kapteyn, and I. P. Christov, "Enhanced high harmonic generation using 25 fs laser pulses," *Physical Review Letters* 76, 752 (1996).

[10] Z. Chang, A. Rundquist, H. Wang, M. M. Murnane, and H. C. Kapteyn, "Generation of coherent soft x-rays at 2.7 nm using high harmonics," *Physical Review Letters* 79, 2967 (1997).

[11] M. Uiberacker, Th. Uphues, M. Schultze, A. J. Verhoef, V. Yakovlev, M. F. Kling, J. Rauschenberger, N. M. Kabachnik, H. Schröder, M. Lezius, K. L. Kompa, H.-G. Muller, M. J. J. Vrakking, S. Hendel, U. Kleineberg, U. Heinzmann, M. Drescher and F. Krausz, "Attosecond real-time observation of electron tunnelling in atoms," *Nature* 446, 627, 2007.

[12] E. Gagnon, P. Ranitovic, X.-M. Tong, C. L. Cocke, M. M. Murnane, H. C. Kapteyn and A. S. Sandhu, "Soft x-ray-driven femtosecond molecular dynamics", *Science* 317, 1347, 2007.

[13] A. L. Cavalieri, N. Müller, Th. Uphues, V. S. Yakovlev, A. Baltuška, B. Horvath, B. Schmidt, L. Blümel, R. Holzwarth, S. Hendel, M. Drescher, U. Kleineberg, P. M. Echenique, R. Kienberger, F. Krausz and U. Heinzmann, "Attosecond spectroscopy in condensed matter," *Nature* 449, 1029, 2007.

[14] W. Theobald, R. Häßner, C. Wülker, and R. Sauerbrey, "Temporally resolved measurement of electron densities ($>10^{23}cm^{-3}$) with high harmonics," *Physical Review Letters* 77, 298 (1996).

[15] R. L. Sandberg, A. Paul, D. A. Raymondson, S. Hädrich, D. M. Gaudiosi, J. Holtsnider, R. I. Tobey, O. Cohen, M. M. Murnane, H. C. Kapteyn, C. Song, J. Miao, Y. Liu, and F. Salmassi, "Lensless diffractive imaging using tabletop coherent high-harmonic soft x-ray beams," *Physical Review Letters* 99, 098103 (2007).

[16] E. A. Gibson, A. Paul, N. Wagner, R. I. Tobey, D. Gaudiosi, S. Backus, I. P. Christov, A. Aquila, E. M. Gullikson, D. T. Attwood, M. M. Murnane, and H. C. Kapteyn, "Coherent soft x-ray generation in the water window with quasi-phase matching," *Science* 302, 95 (2003).

[17] A. Rundquist, C. G. Durfee, Z. Chang, C. Herne, S. Backus, M. M. Murnane, and H. C. Kapteyn, "Phase-matched generation of coherent soft x-rays," *Science* 280, 1412 (1998).

[18] A. L'Huillier and P. Balcou, "High-order harmonic generation in rare gases with a 1-ps 1053-nm laser," *Physical Review Letters* 70, 774 (1993).

[19] C. Winterfeldt, C. Spielmann, and G. Gerber, "Colloquium: Optimal control of high-harmonic generation," *Reviews of Modern Physics* 80, 117 (2008).

[20] M. V. Ammosov, N. B. Delone, and V. P. Krainov, "Tunnel ionization of complex atoms and of atomic ions in an alternating electromagnetic field," *Soviet Physics JETP* 64, 1191 (1986).

[21] M. Lewenstein, P. Balcou, M. Y. Ivanov, A. L'Huillier, and P. B. Corkum, "Theory of high-harmonic generation by low-frequency laser fields," *Physical Review* A49, 2117 (1994).

[22] P. Balcou, and A. L'Huillier, "Phase-matching effects in strong-field harmonic generation," *Physical Review* A 47, 1447 (1993).

[23] J. Peatross, and D. D. Meyerhofer, "Intensity-dependent atomic-phase effects in high-order harmonic generation," *Physical Review A*52, 3976 (1995).

[24] A. L'Huillier, X. F. Li, and L. A. Lompré, "Propagation effects in high-order harmonic generation in rare gases," *Journal of the Optical Society of America* B 7, 527 (1990).

[25] A. L'Huillier, K. J. Schafer and K. C. Kulander, "Theoretical aspects of intense field harmonic generation," *Journal of Physics B: Atomic, Molecular, and Optical Physics* 24, 3315 (1991).

[26] A. L. Lytle, "Phase Matching and Coherence of High Order Harmonic Generation in Hollow Waveguides" *PhD thesis,* University of Colorado (2008)

[27] A. Averchi, D. Faccio, R. Berlasso, M. Kolesik, J. V. Moloney, A. Couairon, and P. Di Trapani, "Phase matching with pulsed Bessel beams for high-order harmonic generation," *Physical Review* A 77, 021802 (2008).

[28] T. Pfeifer, C. Spielmann and G. Gerber, "Femtosecond x-ray science", *Reports on Progress in Physics* 69, 443 (2006).

[29] C. G. Durfee, III, A. R. Rundquist, S. Backus, C. Herne, M. M. Murnane, and H. C. Kapteyn, "Phase matching of high-order harmonics in hollow waveguides," *Physical Review Letters* 83, 2187 (1999).

[30] V. Tosa, E. Takahashi, Y. Nabekawa, and K. Midorikawa, "Generation of high-order harmonics in a self-guided beam," *Physical Review* A 67, 063817 (2003).

[31] P. W. Miloni, and J. H. Eberly, "Lasers", John Wiley and Sons, New York, Wiley Series in *Pure and Applied Optics Edition* (1998).

[32] P. Salières, A. L'Huillier, and M. Lewenstein, "Coherence control of high-order harmonics," *Physical Review Letters* 74, 3776 (1995).

[33] J. T. Verdeyen, and J. B. Gerardo, "Application of the laser to plasma refractive index determination," *Annals of the New York Academy of Science* 122, 676 (1965).

[34] T. Popmintcheva, M. C. Chena, A. Bahabada, M. Gerritya, P. Sidorenkob, O. Cohenb, I. P. Christovc, M. M. Murnanea, and H. C. Kapteyna, "Phase matching of high harmonic generation in the soft and hard X-ray regions of the spectrum," *Proceedings of the National Academy of Sciences of the United States of America* 106, 10516 (2009).

[35] J. Peatross, M. V. Fedorov, and K. C. Kulander, "Intensity-dependent phase-matching effects in harmonic generation," *Journal of the OpticalSociety of America* B 12, 863 (1995).

[36] M. B. Gaarde, F. Salin, E. Constant, P. Balcou, K. J. Schafer, K. C. Kulander, and A. L'Huillier, "Spatiotemporal separation of high harmonic radiation into two quantum path components," *Physical Review* A 59, 1367 (1999).

[37] T. Auguste, B. Carré, and P. Salières, "Quasi-phase-matching of high-order harmonics using a modulated atomic density," *Physical Review*A 76, 011802 (2007).

[38] E. Constant, D. Garzella, P. Breger, E. Mével, C. Dorrer, C. Le Blanc, F. Salin, and P. Agostini, "Optimizing high harmonic generation in absorbing gases: model and experiment," *Physical Review Letters* 82, 1668 (1999).

[39] S. Kazamias, D. Douillet, F. Weihe, C. Valentin, A. Rousse, S. Sebban, G. Grillon, F. Augé, D. Hulin, and P. Balcou, "Global optimization of high harmonic generation," *Physical Review Letters* 90, 193901 (2003).

[40] J. F. Hergott, M. Kovacev, H. Merdji, C. Hubert, Y. Mairesse, E. Jean, P. Breger, P. Agostini, B. Carré, and P. Salières, "Extreme-ultraviolet high-order harmonic pulses in the microjoule range," *Physical Review* A 66, 021801 (2002).

[41] Y. Tamaki, J. Itatani, Y. Nagata, M. Obara, and K. Midorikawa, "Highly efficient, phase-matched high-harmonic generation by a self-guided laser beam," *Physical Review Letters* 82, 1422 (1999).

[42] C. Altucci, R. Bruzzese, D. D'Antuoni, C. de Lisio, and S. Solimeno, "Harmonic generation in gases by use of Bessel-Gauss laser beams," *Journal of the OpticalSociety of America* B 17, 34 (2000).

[43] L. V. Dao, S. Teichmann, J. Davis, and P. Hannaford, "Generation of high flux, highly coherent extreme ultraviolet radiation in a gas cell," *Journal of Applied Physics* 104, 023105 (2008).

[44] L. V. Dao, S. Teichmann, and P. Hannaford, "Phase-matching for generation of few high order harmonics in a semi-infinite gas cell," *Physics Letters* A 372, 5254 (2008).

[45] J. A. Armstrong, N. Bloembergen, J. Ducuing, and P. S. Pershan, "Interactions between light waves in a nonlinear dielectric," *Physical Review* 127, 1918 (1962).

[46] J. Seres, V. S. Yakovlev, E. Seres, C. Streli, P. Wobrauschek, C. Spielmann, and F. Krausz, "Coherent superposition of laser-driven soft-X-ray harmonics from successive sources," *NaturePhysics* 3, 878 (2007).

[47] J. Peatross, S. Voronov, and I. Prokopovich, "Selective zoning of high harmonic emission using counter-propagating light," *Optics Express* 1, 114 (1997).

[48] S. L. Voronov, I. Kohl, J. B. Madsen, J. Simmons, N. Terry, J. Titensor, Q. Wang, and J. Peatross, "Control of laser high-harmonic generation with counterpropagating light," *Physical Review Letters* 87, 133902 (2001).

[49] J. B. Madsen, L. A. Hancock, S. L. Voronov, and J. Peatross, "High-order harmonic generation in crossed laser beams," *Journal of the OpticalSociety of America* B 20, 166 (2003).

[50] Xiaoshi Zhang, Amy L. Lytle, Tenio Popmintchev, Xibin Zhou, Henry C. Kapteyn, Margaret M. Murnane, and Oren Cohen, "Quasi-phase-matching and quantum-path control of high-harmonic generation using counterpropagating light," *Nature Physics* 3, 270 (2007).

[51] S. Teichmann, "High harmonic generation for coherent diffractive imaging," *PhD thesis,* Swinburne University of Technology (2009).

[52] C. F. R. Caron and R. M. Potvliege, "Optimum conical angle of a Bessel-Gauss beam for low-order harmonic generation in gases," *Journal of the OpticalSociety of America* B 16, 1377 (1999).

[53] T. Auguste, O. Gobert, and B. Carré, "Numerical study on high-order harmonic generation by a Bessel-Gauss laser beam," *Physical Review* A 78, 033411 (2008).

[54] M. Nisoli, E. Priori, G. Sansone, S. Stagira, G. Cerullo, S. De Silvestri, C. Altucci, R. Bruzzese, C. de Lisio, P. Villoresi, L. Poletto, M. Pascolini, and G. Tondello, "High-brightness high-order harmonic generation by truncated Bessel beams in the sub-10-fs regime," *Physical Review Letters* 88, 033902 (2002).

[55] S. Kazamias, D. Douillet, C. Valentin, F. Weihe, F. Augé, T. Lefrou, G. Grillon, S. Sebban, and P. Balcou, "Observation of high-contrast coherence fringes in high-order harmonic generation," *Physical Review* A 68, 033819 (2003).

[56] J. P. Brichta, M. C. H. Wong, J. B. Bertrand, H. C. Bandulet, D. M. Rayner, and V. R. Bhardwaj, "Comparison and real-time monitoring of high-order harmonic generation in different sources," *Physical Review* A 79, 033404 (2009).

[57] M. Schnüer, Z. Cheng, M. Hentschel, G. Tempea, P. Kálmán, T. Brabec, and F. Krausz, "Absorption-limited generation of coherent ultrashort soft x-ray pulses," *Physical Review Letters* 83, 722 (1999).

[58] J. Peatross and D. D. Meyerhofer, "Angular distribution of high-order harmonics emitted from rare gases at low density," *Physical Review* A 51, R906 (1995).

[59] E. A. Gibson, A. Paul, N. Wagner, R. Tobey, S. Backus, I. P. Christov, M. M. Murnane, and H. C. Kapteyn, "High-order harmonic generation up to 250 eV from highly ionized argon," *Physical Review Letters* 92, 033001 (2004).

[60] T. Popmintchev, M.-C. Chen, O. Cohen, M. E. Grisham, J. J. Rocca, M. M. Murnane, and H. C. Kapteyn, "Extended phase matching of high harmonics driven by mid-infrared light," *OpticsLetters* 33, 2128 (2008).

[61] J. J. Carrera, and S.-I. Chu, "Extension of high-order harmonic generation cutoff via coherent control of intense few-cycle chirped laser pulses," *Physical Review* A 75, 033807 (2007).

[62] J. Tate, T. Auguste, H. G. Muller, P. Salières, P. Agostini, and L. F. DiMauro, "Scaling of wave-packet dynamics in an intense midinfrared field," *Physical Review Letters* 98, 013901 (2007).

[63] M. Lewenstein, P. Salières, and A. L'Huillier, "Phase of the atomic polarization in high-order harmonic generation," *Physical Review* A 52, 4747 (1995).

[64] P. Antoine, A. L'Huillier, M. Lewenstein, P. Salières, and B. Carré, "Theory of high-order harmonic generation by an elliptically polarized laser field," *Physical Review* A 53, 1725 (1996).

[65] W. M. Wood, C. W. Siders, and M. C. Downer, "Measurement of femtosecond ionization dynamics of atmospheric density gases by spectral blueshifting," *Physical Review Letters* 67, 3523 (1991).

[66] S. C. Rae, and K. Burnett, "Detailed simulations of plasma-induced spectral blueshifting," *Physical Review* A 46, 1084 (1992).

[67] K. Miyazaki, and H. Takada, "High-order harmonic generation in the tunneling regime," *Physical Review* A 52, 3007 (1995).

[68] C. G. Wahlström, J. Larsson, A. Persson, T. Starczewski, S. Svanberg, P. Salières, P. Balcou, and A. L'Huillier, "High-order harmonic

generation in rare gases with an intense short-pulse laser," *Physical Review* A 48, 4709 (1993).

[69] C. Altucci, R. Bruzzese, C. de Lisio, M. Nisoli, S. Stagira, S. De Silvestri, O. Svelto, A. Boscolo, P. Ceccherini, L. Poletto, G. Tondello, and P. Villoresi, "Tunable soft-x-ray radiation by high-order harmonic generation," *Physical Review* A 61, 021801 (1999).

[70] C. Kan, C. E. Capjack, R. Rankin, and N. H. Burnett, "Spectral and temporal structure in high harmonic emission from ionizing atomic gases," *Physical Review* A 52, R4336 (1995).

[71] C. M. Hey, J. Güdde, U. Höfer, and A. L'Huillier, "Spectrally resolved maker fringes in high-order harmonic generation," *Physical Review Letters,* 107, 033903 (2011).

[72] A. Zaïr, M. Holler, A. Guandalini, F. Schapper, J. Biegert, L. Gallmann, U. Keller, A. S. Wyatt, A. Monmayrant, I. A. Walmsley, E. Cormier, T. Auguste, J. P. Caumes, and P. Salières, "Quantum path interferences in high-order harmonic generation," *Physical Review Letters* 100, 143902 (2008).

[73] T. Auguste, P. Salières, A. S. Wyatt, A. Monmayrant, I. A. Walmsley, E. Cormier, A. Zaïr, M. Holler, A. Guandalini, F. Schapper, J. Biegert, L. Gallmann, and U. Keller, "Theoretical and experimental analysis of quantum path interferences in high-order harmonic generation," *Physical Review* A 80, 033817 (2009).

[74] E. Brunetti, R. Issac, and D. A. Jaroszynski, "Quantum path contribution to high-order harmonic spectra," *Physical Review* A 77, 023422 (2008).

[75] F. Zhong, J. Deng, X. Hu, Z. Li, Z. Zhang, and Z. Xu, "The effect of ionization of gases on the high harmonic splitting," *Physics Letters* A 278, 35 (2000).

[76] Z. Fangchuan, L. Zhong, Z. Zhinan, Z. Zhengquan, L. Ruxin, and X. Zhizhan, "Spectral splitting of high-order harmonic emissions from ionizing gases," *Physical Review* A 65, 033808 (2002).

[77] Y. Wang, Y. Liu, X. Yang, and Z. Xu, "Spectral splitting in high-order harmonic generation," *Physical Review* A 62, 063806 (2000).

[78] K. Varjú, Y. Mairesseb, B. Carré, M. B. Gaarde, P. Johnsson, S. Kazamias, R. L. Martens, J. Mauritsson, K. J. Schafer, P. Balcou, A. L'Huillier and P. Salières, "Frequency chirp of harmonic and attosecond pulses," *Journal of Modern Physics,* 52, 379 (2005).

[79] K. J. Schafer, and K. C. Kulander, "High harmonic generation from ultrafast pump lasers," *Physical Review Letters* 78, 638 (1997).

[80] Z. Chang, A. Rundquist, H. Wang, I. Christov, H. C. Kapteyn, and M. M. Murnane, "Temporal phase control of soft-x-ray harmonic emission," *Physical Review* A 58, R30 (1998).

[81] R. Bartels, S. Backus, E. Zeek, L. Misoguti, G. Vdovin, I. P. Christov, M. M. Murnane, and H. C. Kapteyn, "Shaped-pulse optimization of coherent emission of high-harmonic soft X-rays," *Nature* 406, 164 (2000).

[82] D. G. Lee, J.-H. Kim, K.-H. Hong, and C. H. Nam, "Coherent control of high-order harmonics with chirped femtosecond laser pulses," *Physical Review Letters* 87, 243902 (2001).

[83] C. A. Froud, E. T. Rogers, D. C. Hanna, W. S. Brocklesby, M. Praeger, A. M. de Paula, J. J. Baumberg, and J. G. Frey, "Soft-x-ray wavelength shift induced by ionization effects in a capillary," *Optics Letters* 31, 374 (2006).

[84] H. T. Kim, I. J. Kim, K. H. Hong, D. G. Lee, J. H. Kim and C. H. Nam, "Chirp analysis of high-order harmonics from atoms driven by intense femtosecond laser pulses," Journal of Physics B: Atomic, *Molecular and Optical Physics,* 37, 1141 (2004).

[85] P. Balling, D. J. Maas, and L. D. Noordam, "Interference in climbing a quantum ladder system with frequency-chirped laser pulses," *Physical Review* A 50, 4276 (1994).

[86] T. Brabec, and F. Krausz, "Intense few-cycle laser fields: Frontiers of nonlinear optics," *Reviews of Modern Physics* 72, 545 (2000).

[87] E. Treacy, "Optical pulse compression with diffraction gratings," *IEEE Journal of Quantum Electronics,* 5, 454 (1969).

[88] G. D. Reid, K. Wynne, "Ultrafast laser technology and spectroscopy," *Encyclopedia and Analytical chemistry,* John Wiley & Sons Ltd. (2006).

[89] K. H. Hong, J. H. Sung, Y. S. Lee, and C. H. Nam, "Temporal characterization of chirped femtosecond laser pulses," *Optics Communications* 213, 193 (2002).

[90] N. Hay, R. Velotta, M. Lein, R. de Nalda, E. Heesel, M. Castillejo, and J. P. Marangos, "High-order harmonic generation in laser-aligned molecules," *Physical Review* A 65, 053805 (2002).

[91] J. Itatani, J. Levesque, D. Zeidler, H. Niikura, H. Pepin, J. C. Kieffer, P. B. Corkum, and D. M. Villeneuve, "Tomographic imaging of molecular orbitals," *Nature* 432, 867 (2004).

[92] T. Kanai, S. Minemoto, and H. Sakai, "Quantum interference during high-order harmonic generation from aligned molecules," *Nature* 435, 470 (2005).

[93] R. A. Bartels, N. L. Wagner, M. D. Baertschy, J. Wyss, M. M. Murnane, and H. C. Kapteyn, "Phase-matching conditions for nonlinear frequency conversion by use of aligned molecular gases," *Optics Letters* 28, 346 (2003).

[94] S. Varma, Y. H. Chen, and H. M. Milchberg, "Trapping and destruction of long-range high-intensity optical filaments by molecular quantum wakes in air," *Physical Review Letters* 101, 205001 (2008).

[95] B. K. McFarland, J. P. Farrell, P. H. Bucksbaum, and M. Gühr, "High harmonic generation from multiple orbitals in N_2," *Science* 322, 1232 (2008).

[96] S. Haessler, J. Caillat, W. Boutu, C. Giovanetti-Teixeira, T. Ruchon, T. Auguste, Z. Diveki, P. Breger, A. Maquet, B. Carré, R. Taieb, and P. Salières, "Attosecond imaging of molecular electronic wavepackets," *Nature Physics* 6, 200 (2010).

[97] O. Smirnova, Y. Mairesse, S. Patchkovskii, N. Dudovich, D. Villeneuve, P. Corkum, and M. Y. Ivanov, "High harmonic interferometry of multi-electron dynamics in molecules," *Nature* 460, 972 (2009).

[98] M. Morgen, W. Price, P. Ludowise, and Y. Chen, "Tensor analysis of femtosecond Raman induced polarization spectroscopy: Application to the study of rotational coherence," *The Journal of Chemical Physics* 102, 8780 (1995).

[99] H. Stapelfeldt, and T. Seideman, "Colloquium: Aligning molecules with strong laser pulses," *Reviews of Modern Physics* 75, 543 (2003).

[100] E. T. J. Nibbering, G. Grillon, M. A. Franco, B. S. Prade, and A. Mysyrowicz, "Determination of the inertial contribution to the nonlinear refractive index of air, N_2, and O_2 by use of unfocused high-intensity femtosecond laser pulses," *Journal of the OpticalSociety of America* B 14, 650 (1997).

[101] J. F. Ripoche, G. Grillon, B. Prade, M. Franco, E. Nibbering, R. Lange, A. Mysyrowicz, "Determination of the time dependence of N_2 in air," *Optics Communications* 135, 310 (1997).

[102] B. Friedrich, and D. Herschbach, "Alignment and trapping of molecules in intense laser fields," *Physical Review Letters* 74, 4623 (1995).

[103] Y. Gao, C. Wu, N. Xu, G. Zeng, H. Jiang, H. Yang, and Q. Gong, "Manipulating molecular rotational wave packets with strong femtosecond laser pulses," *Physical Review* A 77, 043404 (2008).

[104] P. W. Dooley, I. V. Litvinyuk, K. F. Lee, D. M. Rayner, M. Spanner, D. M. Villeneuve, and P. B. Corkum, "Direct imaging of rotational wave-

packet dynamics of diatomic molecules," *Physical Review* A 68, 023406 (2003).

[105] F. Rosca-Pruna, and M. J. J. Vrakking, "Experimental observation of revival structures in picosecond laser-induced alignment of I_2," *Physical Review Letters* 87, 153902 (2001).

[106] C. Horn, M. Wollenhaupt, M. Krug, T. Baumert, R. de Nalda, and L. Bañares, "Adaptive control of molecular alignment," *Physical Review*A 73, 031401 (2006).

[107] M. Renard, E. Hertz, S. Guérin, H. R. Jauslin, B. Lavorel, and O. Faucher, "Control of field-free molecular alignment by phase-shaped laser pulses," *Physical Review* A 72, 025401 (2005).

[108] V. Renard, M. Renard, S. Guérin, Y. T. Pashayan, B. Lavorel, O. Faucher, and H. R. Jauslin, "Postpulse molecular alignment measured by a weak field polarization technique," *Physical ReviewLetters* 90, 153601 (2003).

[109] J. Itatani, D. Zeidler, J. Levesque, M. Spanner, D. M. Villeneuve, and P. B. Corkum, "Controlling high harmonic generation with molecular wave packets," *Physical Review Letters* 94, 123902 (2005).

[110] H. Cai, J. Wu, A. Couairon, and H. Zeng, "Spectral modulation of femtosecond laser pulse induced by molecular alignment revivals," *Optics Letters* 34, 827 (2009).

[111] V. Kumarappan, C. Z. Bisgaard, S. S. Viftrup, L. Holmegaard, and H. Stapelfeldt, "Role of rotational temperature in adiabatic molecular alignment," *The Journal of Chemical Physics,* 125, 194309 (2006).

[112] K. Yoshii, G. Miyaji, and K. Miyazaki, "Measurement of molecular rotational temperature in a supersonic gas jet with high-order harmonic generation," *Optics Letters* 34, 1651 (2009).

[113] N. Xu, C. Wu, Y. Gao, H. Jiang, H. Yang, and Q. Gong, "Measurement of the field-free alignment of diatomic molecules," *The Journal of Physical Chemistry* A 112, 612 (2008).

[114] I. V. Litvinyuk, K. F. Lee, P. W. Dooley, D. M. Rayner, D. M. Villeneuve, and P. B. Corkum, "Alignment-dependent strong field ionization of molecules," *Physical Review Letters* 90, 233003 (2003).

[115] X. M. Tong, Z. X. Zhao, and C. D. Lin, "Theory of molecular tunneling ionization," *Physical Review*A 66, 033402 (2002).

[116] J. Muth-Böhm, A. Becker, and F. H. M. Faisal, "Suppressed molecular ionization for a class of diatomics in intense femtosecond laser fields," *Physical Review Letters* 85, 2280 (2000).

[117] G. H. Lee, I. J. Kim, S. B. Park, T. K. Kim, Y. S. Lee and C. H. Nam, "Alignment dependence of high harmonics contributed from HOMO and HOMO-1 orbitals of N_2 molecules," Journal of Physics B: Atomic, *Molecular and Optical Physics* 43, 205602 (2010).

[118] A. D. Bandrauk, and N. H. Shon, "Attosecond control of ionization and high-order harmonic generation in molecules," *Physical Review* A 66, 031401 (2002).

[119] B. A. Sickmiller, and R. R. Jones, "Effects of phase-matching on high-order harmonic generation from aligned N_2," *Physical Review* A 80, 031802 (2009).

[120] V. Loriot, E. Hertz, O. Faucher, and B. Lavorel, "Measurement of high order Kerr refractive index of major air components," *Optics Express* 17, 13429 (2009).

[121] Z. X. Zhao, X. M. Tong, and C. D. Lin, "Alignment-dependent ionization probability of molecules in a double-pulse laser field," *Physical Review* A 67, 043404 (2003).

[122] F. J. Giessibl, "Advances in atomic force microscopy," *Reviews of Modern Physics* 75, 949-983 (2003).

[123] J. C. H. Spence, *"High resolution electron microscopy,"* Oxford University Press, 3rd edition (2009).

[124] C. Jacobsen, "Soft x-ray microscopy," *Trends in Cell Biology,* 9, 44 (1999).

[125] S. M. Hurtley, and L. Helmuth, "The future looks bright," *Science* 300, 75 (2003).

[126] D. Attwood, *"Soft x-rays and extreme ultraviolet radiation,"* Cambridge University Press, 1st edition (1999).

[127] W. Chao, B. D. Harteneck, J. A. Liddle, E. H. Anderson, and D. T. Attwood, "Soft X-ray microscopy at a spatial resolution better than 15 nm," *Nature* 435, 1210 (2005).

[128] I. McNulty, J. Kirz, C. Jacobsen, E. H. Anderson, M. R. Howells, and D. P. Kern, "High-resolution imaging by Fourier transform x-ray holography," *Science* 256, 1009 (1992).

[129] J. Miao, P. Charalambous, J. Kirz, and D. Sayre, "Extending the methodology of X-ray crystallography to allow imaging of micrometre-sized non-crystalline specimens," *Nature* 400, 342 (1999).

[130] I. K. Robinson, I. A. Vartanyants, G. J. Williams, M. A. Pfeifer, and J. A. Pitney, "Reconstruction of the shapes of gold nanocrystals using coherent x-ray diffraction," *Physical Review Letters* 87, 195505 (2001).

[131] G. J. Williams, H. M. Quiney, A. G. Peele, and K. A. Nugent, "Coherent diffractive imaging and partial coherence," *Physical Review* B 75, 104102 (2007).

[132] M. A. Pfeifer, G. J. Williams, I. A. Vartanyants, R. Harder, and I. K. Robinson, "Three-dimensional mapping of a deformation field inside a nanocrystal," *Nature* 442, 63 (2006).

[133] G. J. Williams, H. M. Quiney, B. B. Dhal, C. Q. Tran, K. A. Nugent, A. G. Peele, D. Paterson, and M. D. de Jonge, "Fresnel coherent diffractive imaging," *Physical Review Letters* 97, 025506 (2006).

[134] S. Marchesini, H. He, H. N. Chapman, S. P. Hau-Riege, A. Noy, M. R. Howells, U. Weierstall, and J. C. H. Spence, "X-ray image reconstruction from a diffraction pattern alone," *Physical Review* B 68, 140101 (2003).

[135] J. Miao, T. Ishikawa, B. Johnson, E. H. Anderson, B. Lai, and K. O. Hodgson, "High resolution 3D x-ray diffraction microscopy," *Physical Review Letters* 89, 088303 (2002).

[136] C. G. Schroer, P. Boye, J. M. Feldkamp, J. Patommel, A. Schropp, A. Schwab, S. Stephan, M. Burghammer, S. Schöder, and C. Riekel, "Coherent x-ray diffraction imaging with nanofocused illumination," *Physical Review Letters* 101, 090801 (2008).

[137] D. Shapiro, P. Thibault, T. Beetz, V. Elser, M. Howells, C. Jacobsen, J. Kirz, E. Lima, H. Miao, A. M. Neiman, and D. Sayre, "Biological imaging by soft x-ray diffraction microscopy," *Proceedings of the National Academy of Sciences of the United States of America* 102, 15343 (2005).

[138] H. N. Chapman, A. Barty, M. J. Bogan, S. Boutet, M. Frank, S. P. Hau-Riege, S. Marchesini, B. W. Woods, S. Bajt, W. H. Benner, R. A. London, E. Plonjes, M. Kuhlmann, R. Treusch, S. Dusterer, T. Tschentscher, J. R. Schneider, E. Spiller, T. Moller, C. Bostedt, M. Hoener, D. A. Shapiro, K. O. Hodgson, D. van der Spoel, F. Burmeister, M. Bergh, C. Caleman, G. Huldt, M. M. Seibert, F. R. N. C. Maia, R. W. Lee, A. Szoke, N. Timneanu, and J. Hajdu, "Femtosecond diffractive imaging with a soft-X-ray free-electron laser," *Nature Physics* 2, 839 (2006).

[139] H. N. Chapman, A. Barty, S. Marchesini, A. Noy, S. P. Hau-Riege, C. Cui, M. R. Howells, R. Rosen, H. He, J. C. H. Spence, U. Weierstall, T. Beetz, C. Jacobsen, and D. Shapiro, "High-resolution ab initio three-dimensional x-ray diffraction microscopy," *Journal of the OpticalSociety of America* A, 1179 (2006).

[140] R. A. Dilanian, B. Chen, G. J. Williams, H. M. Quiney, K. A. Nugent, S. Teichmann, P. Hannaford, L. V. Dao, and A. G. Peele, "Diffractive imaging using a polychromatic high-harmonic generation soft-x-ray source," *Journal of Applied Physics* 106, 023110 (2009).

[141] B. Chen, R. A. Dilanian, S. Teichmann, B. Abbey, A. G. Peele, G. J. Williams, P. Hannaford, L. V. Dao, H. M. Quiney, and K. A. Nugent, "Multiple wavelength diffractive imaging," *Physical Review* A 79, 023809 (2009).

[142] S. Teichmann, B. Chen, R. A. Dilanian, P. Hannaford, and L. V. Dao, "Experimental aspects of multiharmonic-order coherent diffractive imaging," *Journal of Applied Physics* 108, 023106 (2010).

[143] J. Miao and D. Sayre, "On possible extensions of x-ray crystallography throughdiffraction-pattern oversampling," *Acta Crystallographica* A, 56 (2000).

[144] J. Miao, T. Ishikawa, E. H. Anderson, and K. O. Hodgson, "Phase retrieval of diffraction patterns from noncrystalline samples using the oversampling method," *Physical Review* B 67, 174104 (2003).

[145] D. Sayre, "Some implications of a theorem due to Shannon," *Acta Crystallographica* 5, 843 (1952).

[146] J. Miao, D. Sayre and H. N. Chapman, "Phase retrieval from the magnitude of the Fourier transforms of nonperiodic objects," *Journal of the OpticalSociety of America* A 15, 1662 (1998).

[147] J. R. Fienup, "Phase retrieval algorithms: a comparison," *Applied Optics* 21, 2758 (1982).

[148] R. A. Dilanian, B. Chen, S. Teichmann, L. V. Dao, H. M. Quiney, and K. A. Nugent, "High-harmonic-generation spectrum reconstruction from Young's double-slits interference pattern using the maximum entropy method," *Optics Letters* 33, 2341 (2008).

[149] S. Teichmann, B. Chen, R. A. Dilanian, P. Hannaford and L. V. Dao, "Spectral characteristics across the spatial profile of a high harmonic beam," *Journal of Physics D: Applied Physics* 42, 135108 (2009).

[150] F. Izumi, and R. A. Dilanian, "Structure refinement based on the maximum entropy method from powder diffraction data," *Recent Research Developments in Physics* 2, 699 (2002).

INDEX

D

E

F

G

H

O

P

Q

R

S

T

U

V

W

X

Y

Z